FEEDING A HUNGRY PLANET

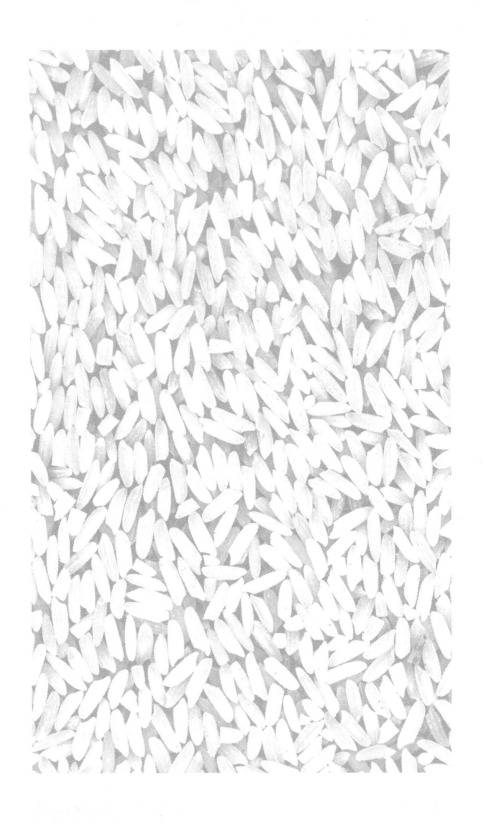

JAMES LANG

FEEDING A HUNGRY PLANET

RICE, RESEARCH, & DEVELOPMENT IN ASIA & LATIN AMERICA

The University of North Carolina Press / Chapel Hill & London

Manufactured in the United States of America

The paper in this book meets the guidelines for permanence and durability of the Committee on Production Guidelines for Book Longevity of the Council on Library Resources.

Library of Congress Cataloging-in-Publication Data
Lang, James, 1944–
Feeding a hungry planet: rice, research, and development in Asia and Latin America / James Lang.
p. cm.
Includes bibliographical references (p.) and index.
ISBN 0-8078-2284-1 (cloth: alk. paper).
ISBN 0-8078-4593-0 (pbk.: alk. paper)
1. Rice—Asia. 2. Rice—South America. 3. Rice—Varieties—Asia.
4. Rice—Varieties—South America. 5. Agricultural innovations—Asia.
6. Agricultural innovations—South America.
7. Rice—Research—Asia.
8. Rice— Research—South America. I. Title.
SB191.R5L37 1996
338.1′6—dc20 95-50150
CIP

00 99 98 97 96 5 4 3 2 1

ACKNOWLEDGMENTS

This book is not mine. It belongs to the scientists, extension workers, and farmers who feed this hungry planet. What is original belongs to them. The mistakes are mine.

With support from the Rockefeller Foundation, I visited four international agricultural research centers: the International Center for Tropical Agriculture, the International Potato Center, the International Maize and Wheat Improvement Center, and the International Rice Research Institute. I also went to local projects and research stations in Brazil, Colombia, Chile, and the Philippines. I promised the Rockefeller Foundation a book about agriculture people would actually read. I promised them the details and the big picture. I hope I kept my promise.

I am grateful to Vanderbilt University and the Department of Sociology for their support. The dean of the college, Jacque Voegeli, and my chairman, Jack Gibbs, approved a leave staggered over two years. I did most of the fieldwork during the spring and summer of 1989 and 1990.

The Rockefeller Foundation did not impose its views on my work. It is certainly not responsible for my conclusions; I went to the Rockefeller Foundation only once in my life.

Jack Reeves encouraged me to keep writing and doing fieldwork. In 1993, I went to the Asian Vegetable Research and Development Center in Taiwan; in 1994, to the International Crops Research Institute for the Semi-Arid Tropics in India; and in 1995, to the International Potato Center's regional office in Indonesia. Thank you Jack.

Charles Tilly encouraged me to do what I wanted to do. Thank you Chuck. William A. Christian read, marked up, and complained about the original version of the book. Thank you Bill. Robert Huggan at IRRI and Tom Hargrove at CIAT kept me supplied with reports, newsletters, and press releases; they answered my questions; they gave me good advice.

I am grateful for the reviews by Tom Hargrove and Alan Fletcher. Karen Moldenhauer at the rice extension center in Stuttgart, Arkansas, checked for rice-specific errors.

At the University of North Carolina Press, I owe a special debt to board member Major Goodman. His pointed criticism, attention to detail, and sound advice gave my editing a sense of direction.

Lew Bateman supported the project from the start and wrote to the Rockefeller Foundation on my behalf. Thank you, Lew.

INTRODUCTION

The rice paddy and the office building create different modes of thought and consciousness. The modern world sees through the eyes of the city. Urban demand for energy, raw materials, and food are at the heart of the planet's ecological crisis. For the city, energy grids, trade, and transit come first; protecting farms, forests, and fragile ecosystems is secondary. Sophisticated urbanites may understand the stock market, but they know next to nothing about the planet's food crops. Understanding a rice paddy, however, is as important for our common future as electronic mail and free trade.

The industries of the future will be knowledge-based: brainpower linked by computers and fiber-optic cables. How easy it is to forget about the knowledge packed into a grain of rice. Agriculture has always been a repository of knowledge: about crop rotations and when to rest the soil, about insect cycles, about reading the signs in nature. It still is. The new global technology of fax machines and the Internet is remarkable, but there is a greater wonder in a rice paddy.

This book is about rice, the crop the world depends on most. It describes how farmers grow it, the distinct environments within which it is produced, and the impact modern varieties have had on yields. It looks at how the knowledge that sustains the crop is preserved internationally, shared by rice-producing countries, and applied locally. It examines ways to grow more rice in a sustainable way and considers prospects for the future.

Between 1960 and 1990, Asia's population nearly doubled, increasing from 1.7 to 3.1 billion.[1] Despite predictions to the contrary, rice production kept ahead of population growth.[2] For productivity, nothing comes close to the earth itself. From just 90 kilos (1 kilo = 2.2 pounds) of seed, small farmers in Asia produce 6 metric tons (1 metric ton = 1,000 kilos) of rice—with nothing more than a mixture of nitrogen, water, earth, and air.

DEFENSE

To look after the computer's welfare, we depend on the corporate world: from IBM and Zenith to Nippon Electric and Phillips. But what about the welfare of the planet's food crops, its rice and wheat, its maize (corn) and potatoes? Whose job is it to safeguard a crop's genetic diversity, track the spread of new insect biotypes, or identify plant diseases? Whose job is it to help the planet feed itself without destroying the natural resources that make food production possible?

During the 1950s, the population of most developing countries increased at unprecedented rates. To feed their growing populations, countries had to produce more food. In the United States, research at land-grant colleges had already improved basic crops such as wheat and maize. Well-organized county-level extension programs promoted new varieties and production methods. Maize is an outstanding example. Average U.S. yields increased from 1.8 metric tons per hectare (1 hectare = 2.47 acres) for 1938–40 to 2.4 tons for 1948–50. A decade later, in 1958–60, yields were up another metric ton. By 1978–80, average maize yields surpassed 6 metric tons.[3]

WHEAT

In 1943, the Rockefeller Foundation and Mexico's Ministry of Agriculture signed a joint agreement for research on food production. The Mexican project followed the research-training-extension model that was so successful in American agriculture. Norman Borlaug, working with his Mexican counterparts, headed up wheat breeding.[4] Success first came in 1955 after a decade of work. Borlaug and his colleagues crossed semidwarf winter wheats with Mexican spring wheats. Orville A. Vogel had developed the semidwarfs at Washington State University's agricultural station in Pullman. The wheat's short stature came from Norin 10, a wheat variety native to Japan. The result for Mexico was a high-yielding wheat that set more grain; it had strong, short stems that supported the added weight without falling over. It took another seven years to overcome susceptibility to rust and to improve the grain quality. In the meantime, improved varieties began reaching farmers and wheat yields started to rise: from an average of less than 1 metric ton per hectare for 1952–53, Mexico's wheat yields increased to 1.5 metric tons for 1957–60. Beginning in 1962, the project released a new line of disease-resistant semidwarfs. For 1967–70, yields averaged 2.8 tons.[5]

To get maximum benefit from improved varieties, Mexican farmers had to buy high-quality seed, apply nitrogen fertilizer, and irrigate the crop. Using this new approach, Mexico's wheat production mounted steadily from 1.4 million metric tons in 1960 to 2.2 million tons in 1970. For 1987–90, wheat production averaged 4 million metric tons with yields of more than 4 metric tons per hectare.[6]

The new semidwarfs produced high yields almost everywhere Mexico produced wheat. As news of the project's success spread, scientists in wheat-producing Asian countries asked to try the new varieties. With support from the Rockefeller Foundation, Borlaug set up wheat trials in India and Pakistan, which agronomists from each country managed. The semidwarf wheats did much better than expected. Between 1966 and 1968, wheat yields in Pakistan rose from .7 metric tons per hectare to 1.1 tons; total wheat production increased from 3.9 million to 6.4 million metric tons. India also had impressive gains in production: from a total of about 11 million metric tons of wheat in 1966 to 16.5 million tons in 1968.[7] The keys to success were the semidwarf wheats and the new production technology. For his work on behalf of the world's food production, Norman Borlaug received the Nobel Prize for Peace in 1970.

RICE

Compared to wheat, research on semidwarf rices lagged dangerously behind.[8] In tropical Asia, rice is a necessity of life. Yet despite growing populations and mounting food deficits, yields improved little. The Rockefeller Foundation wanted to organize an international effort in rice but lacked the capital to support such a venture alone. In 1958, however, the Ford Foundation expressed interest in a joint project. Together the two foundations established the International Rice Research Institute (IRRI) in 1960 in the Philippines. The Ford Foundation provided more than $6 million to build and equip the research facility, and the Rockefeller Foundation hired the staff and paid operating costs. By 1962 IRRI's rice research was underway. The institute's objective was to develop a short, high-yielding, nitrogen-responsive rice variety and to work out a new production system. Three years later, IRRI released its first improved rice variety, IR8.[9]

The Rockefeller Foundation appointed Robert F. Chandler as IRRI's first director general. Chandler, in turn, selected IRRI's first board of trustees. Thereafter, the board appointed new trustees and selected a new director general when the post became vacant. The first board had ten members, including Chandler, ex officio; one representative each from the Rockefeller and Ford foundations; three representatives from the host country, the Philippines; and four at-large members—in this case, distinguished scientists from Japan, Taiwan, India, and Thailand. Terms were staggered. In 1965, the number of at-large trustees was increased to eight. During IRRI's first twenty years, thirty-seven at-large trustees from nineteen countries served on the board. Six were from the United States.[10] The board has played a key role. Its first chairman was J. George Harrar of the Rockefeller Foundation. He was followed in 1963 by Forest F. Hill from the Ford Foundation, who held the post for the next fifteen years. Harrar and Hill, along with Chandler, were IRRI's founders.

Meanwhile, the research the Rockefeller Foundation supported in Mexico was reorganized and expanded. In 1966 the foundation established the International Maize and Wheat Improvement Center (CIMMYT) to work on production problems in developing countries worldwide.[11] The work of the two centers, and the new wheat and rice varieties they created, generated a "green revolution" in agriculture. In just fifteen years, between 1965 and 1980, combined wheat and rice production in developing countries increased by an average of almost 75 percent.[12] India is a prime example. Instead of the famine experts had predicted for the 1970s, the country ended up with a grain surplus. Comparing the periods 1964–65 and 1969–70, average wheat production grew from 11 million to almost 22 million metric tons.[13] The impact on rice was slower but no less significant. Harvests rose from an annual 35 million metric tons in 1965–66 to almost 40 million tons a decade later.[14] In 1970, the United Nations Economic and Social Council awarded its Science Prize jointly to IRRI and CIMMYT.[15]

EXPANSION

Encouraged by their success in wheat and rice, the foundations funded two more centers. The International Center for Tropical Agriculture (CIAT), founded in 1967 near Cali, Colombia, worked on beans and the tropical root crop cassava (maniac). Both crops originated in the Americas but had spread worldwide.[16] The center's mandate also included tropical pastures and rice. As to pastures, the task was to reclaim degraded ones and make new grazing lands ecologically sustainable. With respect to rice, the job was to adapt the Asian semidwarf varieties to Latin America's growing conditions and market preferences. The same year, the International Institute of Tropical Agriculture (IITA) was founded in Ibadan, Nigeria. Its mandate included traditional African crops such as yams, plantains, and cowpeas, as well as maize and soybeans, crops with great potential for food production.

For 1968–69, the combined budget for the four centers came to $3 million—$750,000 to each center by each foundation. In the long run, that was more than private foundations could afford.[17]

With backing from the World Bank, the Food and Agricultural Organization (FAO), and the United Nations Development Program (UNDP), a loose association of foundations, donor countries, and international organizations was formed in 1971. Called the Consultative Group on International Agricultural Research (CGIAR), it took over responsibility for funding the centers. Assisted by the Technical Advisory Committee, CGIAR oversees the work the centers do. That task is considerable.

What started out in Mexico as a small research project in wheat is now a multifaceted strategic operation that is international in scope and regional in character. In 1972 there were six centers, in 1976 there were eleven, and from 1979 to 1991, there were thirteen. In 1995, CGIAR funded sixteen centers.[18]

STRATEGY

Without the added wheat and rice the new semidwarfs made possible, Asia would have faced a terrible food crisis. Instead, many Asian countries, including India, became self-sufficient. In 1980, the added output from improved wheat and rice varieties worldwide was worth an estimated $56 billion.[19]

The transformation the green revolution engendered has, in turn, created new problems. Farmers use much more fertilizer now than before, they spray pesticides even when insects pose no threat, and the genetic base of crops is narrower today than it used to be. The link between IRRI's research, national programs, and local extension is often weak. Nonetheless, supporting rice research is a good investment in the future. At its best, IRRI's work shows how an international, science-based strategy can be geared to the problems of poor countries and small farmers.

To illustrate what the work of a center is like, this book uses IRRI and CIAT as examples. The many factors that can effect crops are illustrated using the case of rice, with examples drawn from countries in Asia and South America. It could have been other crops, other centers, and other countries; I did not lack alternatives.

FIELDWORK

My first visit to a center was to CIAT in 1983. At the time, I was writing about rural development projects in Colombia.[20] What interested me was whether the research done at CIAT actually benefited local farmers. The more fieldwork I did, the more complicated the question became. No matter how good CIAT's research is, it does not set the region's agrarian policy. How quickly a new variety or production strategy spreads depends on a country's approach to research and extension.

A national program's research on a specific crop reflects that crop's importance domestically. Brazil, for example, is Latin America's biggest rice producer. Per capita consumption far exceeds that for Mexico or Bolivia; so too does its investment in rice research. In Goiás State, Brazil has a national rice research center. Farther south, the country's top rice-producing state, Rio Grande do Sul, has its own rice research program and extension system. So Brazil can cooperate with CIAT on rice. To determine how rice research actually works, however, interviews at CIAT were not enough. I had to follow the story to local testing sites and extension outposts in Brazil.

Besides rice, I also did fieldwork at CIAT on cassava, beans, and tropical pastures. But I spent as much time with agronomists at national centers and local research stations as I did at CIAT. And the project soon spread to other centers. It was hard to tell whether my experience at CIAT was typical—after all, there were eight crop-oriented centers. So I added the International Potato Center (CIP) to the scheme and visited potato projects in Chile, Bolivia, and Colombia. Since the green revolution had started with wheat in Mexico, I went to CIMMYT. As was true with other centers and crops, I traced its research back to national programs and local extension, in this case, to Brazil's center for wheat breeding in Rio Grande do Sul and its center for work on maize in Minas Gerais State. In short, there were many centers, many countries, many crops, and no easy answers.

My experience in Latin America did not prepare me for Asia and its rice. A billion people in Asia eat rice every day, at every meal, almost exclusively. For the rest of Asia, including prosperous, urban countries such as Japan and Korea, rice is still a mainstay. At IRRI, I studied the diverse rice-production systems of monsoon Asia. Field trips were mostly to irrigated fields of transplanted paddy rice, the farming system under which most of Asia's rice is produced.

REPORTS

By the time I put a temporary halt to more fieldwork, I had visited extension projects and research centers in eight countries on three continents. I had interviewed farmers, extension workers, and research scientists on topics ranging from breeding strategies, biotechnology, and disease resistance to cropping systems, pest tolerance, and soil stresses. I had worked on many crops, including rice, wheat, maize, potatoes, beans, and cassava. I had typed, arranged, and outlined more than a thousand pages of field notes.

There is much good news to report: CIAT's work on the biological control of cassava pests, how Brazil became self-sufficient in wheat, how CIP helped Andean communities improve potato storage, about resistance in beans to bruchid storage pests, and about sustainable pastures. There are reports about how CIMMYT helped villages produce their own maize seed, about IRRI and what made the rice revolution possible, about how centers preserved a crop's genetic diversity. I did not want to leave out anything.

It took a while to temper the ambition to write one big book. I decided instead to plow my fields one at a time. I had done more work on rice than on any other crop and pursued it in both Latin America and Asia. As a result, this book is mostly about rice. Nonetheless, the problems that beset rice production, from insect pests and acidic soils to diseases and droughts, crosscut crops. There is much detail; but in agriculture, it is the details that often matter the most. With insects it is the specifics about plant hoppers and spittlebugs that count. For agriculture, the biggest story is often, literally, the smallest seed. The particulars help us understand the abstract principles.

In agriculture, IRRI and CIAT recognize there are no definitive solutions. A variety's resistance to disease and pests is frequently temporary rather than permanent. There is always another virus or leaf hopper lurking. Pathogens and insects constantly beat the latest technology. Consequently, security rests on collective effort, from the work of a center's research scientists to national testing programs, extension teams, and farmers. A good way to understand this is to look at rice in detail.

SOURCES

The text that follows is based mostly on interviews in the field with research scientists, agronomists, and extension workers. They told me about the work they did and the problems they faced. Many interviews lasted just an hour; some went on for several days as part of a field trip. In Latin America, I took handwritten notes in Spanish, Portuguese, or English. Later, I used a typewriter to reconstruct a fuller version. My rule was to finish a complete version before leaving a site, so there was usually a chance to go over points that needed clarification. Nonetheless, the notes I ended up with are my version of what people told me. I have kept direct quotes to a minimum, paraphrasing, occasionally with the aid of footnotes, instead. The field notes cited are deposited in the library at Vanderbilt University. They are organized by center, by the national program involved, and by local agencies. Pages are numbered sequentially.

From centers, national agencies, and projects, I returned with many boxes of books, bulletins, and annual reports. Where possible, I have cited research relevant to the topic discussed published by the person interviewed. The book also relies on statistical information published by international, national, and local agencies.[21]

FEEDING A HUNGRY PLANET

ASIA

1

"If I skip rice at breakfast," said Tom Tengco, mounding a sticky heap on his plate, "my stomach grumbles rice, rice, rice until noon." On our way through Manila, we had stopped at a Jolibee, a Filipino-style fast-food joint. It was 6:00 A.M. and the place was crowded. At Jolibee, customers take a break from the chaotic world of jeepneys, trucks, and buses that makes up Manila's morning traffic. Breakfast choices include rice with fish, rice with chicken, and rice with vegetables. Tom added fish to his rice. I asked for french fries, as the potato was the only thing that to me even vaguely suggested breakfast. It took a while for the french frier to warm up.

Tom is a wiry man in his early thirties with a narrow face and intense brown eyes. He works at the International Rice Research Institute (IRRI), located at Los Baños, about 40 miles south of Manila. He has a degree in agronomy from the University of the Philippines, whose main campus adjoins IRRI. An autonomous research and training center, IRRI helped develop the semidwarf rice varieties that revolutionized Asian production. When I met Tom Tengco, I was at IRRI learning about its rice research. I had done this kind of work before at international centers in Colombia, Peru, and Mexico, but none of the interviews, field trips, and stops at local projects had prepared me for rice in Asia.

RICE FACTS

Wheat, rice, and maize (corn) directly supply half the calories consumed by the 5.6 billion people on the earth.[1] Wheat has the edge in acreage. During 1992–94, farmers planted an average of 222 million hectares (1 hectare = 2.47 acres) of wheat, compared to 147 million hectares worldwide for rice and 130 million for maize. As a food crop, however, rice surpasses wheat. On no other crop do so many people depend so much. For 1992–94, the world's harvest of rough rice averaged 530 million metric tons (1 metric ton = 1,000 kilos). Of this, 91 percent was produced in Asia, and Asia has more than half the world's population. In 1994, the three top rice-growing nations, China, India, and Indonesia, had a combined population in excess of two billion; they accounted for almost two-thirds of the 535 million tons of rice the world produced.[2]

About 60 percent of the world's wheat crop goes to human consumption; about a third is exported. By contrast, human consumption accounts for 85 percent of total rice production. During 1991–93, not more than 3 percent of the world's rice crop was traded internationally; in fact, most rice is consumed within ten miles of where it is produced.[3]

How much rice do people eat? In the United States, per capita consumption of milled rice averages about 8 kilos (1 kilo = 2.2 pounds). In China, by contrast, rice consumption is 90 kilos per person. For Indonesia, it is 136 kilos; in Burma and Laos, it is almost 200 kilos. At least a third of the calories for some 2.8 billion Asians comes exclusively from rice; for Bangladesh and most of Southeast Asia, it is more than 70 percent.[4] In Asia, rice is not just another crop; planting, harvesting, and eating rice is a way of life.

Most of Asia's rice is produced by small farmers with irrigated paddies of 1 hectare or less. The Philippines is no exception.[5]

Growing Rice

Tom Tengco's work takes him to many project sites in the Philippines. Getting farmers to try an unfamiliar technology can be difficult. It is one thing to measure how much nitrogen a ground cover "green manure" like *Sesbania rostrata* fixes per hectare, quite another to get farmers to rotate the plant into their rice fields. Not content with ideal results from a research station, IRRI has to know what farmers think about a new rice variety, production practice, or technology. For in the risky world of farmers, the rains come late, high winds damage plants, and fertilizer costs too much or arrives too late.

At Santa Barbara and Guimba, a different planting system, new rotations, and sesbania incorporation are under review. The Philippine Ministry of Agriculture, IRRI, and local farmers work together on the projects. To get to the sites, we first traversed Luzon Island's central valley. It stretches from Manila Bay almost two hundred miles north-northwest to Dagupan on Lingayen Bay. Set off by rolling hills and mountains to the east and west, the valley has rich, fertile soils and water for irrigation; it is the country's most important rice-producing region.

The end of July is the height of the monsoon season and the worst time in the Philippines for a vacation. Every cloud unleashes a torrent of rain, making the sturdiest umbrella useless. For rice, however, my timing was good. Rice is an emergent, water-loving plant. When the monsoons hit, it is time to prepare the fields and transplant rice seedlings into the paddies.

Where the valley flatlands end, terraces begin, ascending gradually and gracefully up the hillsides in unbroken chains. Water for irrigation comes from streams diverted higher up and then fed to the fields. We were not far from Tarlac, the heart of Luzon's rice country. Rice plots, surrounded by mounded earthen dikes, or "bunds," spread out in all directions. Farmers were out with their plows and harrows, "puddling" the muddy mixture of soil and water that transplanted rice needs. For puddling, many farmers still rely on the Philippine water buffalo, or carabao. This beast of burden works in the fields but provides milk and, eventually, meat for the household. And it is easy to maintain. A water buffalo can subsist on grazing and rice straw.[6] Graceful, gentle beasts of great bulk, I admire the dignified pace at which they work. Late in the day, unyoked and off duty, they relax in the irrigation ditches, neck-deep in water.

Each carabao pulls a rectangular "comb" harrow. As it scrapes through the watery mud, it loosens the soil and dissolves the clumps. A harrow has a single row of spiked prongs. Made of wood and iron, it has no movable parts, is a meter wide (1 meter = 39.37 inches), and weighs only about 10 kilos. Farmers carry their harrows from field to field.[7]

A field is usually harrowed twice, making it a pasty, watery muck, soft enough for a delicate rice seedling to be inserted easily. Because it was planting time, people were bent over in the fields transplanting rice seedlings, which they pulled from neatly tied bundles.

In central Luzon, light rains start in early May and mount steadily into July; the heavy monsoon season hits by the month's end. August is the wettest month. Santa Barbara's August tally averages almost 600 millimeters, as much rain as North Dakota's wheat gets in an entire growing season. So much rain falls so fast that diked fields in Santa Barbara fill up with water, even without irrigation. Heavy rains continue through September, decline sharply in October, and end altogether in late November or early December. The dry season sets in by Christmas and continues through most of April.

Farmers in Santa Barbara establish seedbeds during May, after they accumulate enough rainwater to flood a small plot of perhaps 15 square meters. They soak about 90 kilos of rice seed, which they then broadcast densely over the muddy surface of the soil. As the seedlings germinate and the stems shoot up, they increase the water level in the plot. A paddy flooded to a shallow depth of 50 to 100 millimeters (100 millimeters = 4 inches) is ideal. Farmers begin transplanting to their main fields in late July or early August for a harvest in late October or early November. Overcast skies are much reduced by the end of September, allowing increased luminosity from the sun to ripen the grain.[8]

Transplanting a hectare of rice is a tremendous amount of work. First, the soil has to be puddled. Then the seedlings have to be separated into bundles and carried to the fields. To transplant a hectare with rows spaced 20 centimeters (10 centimeters = 4 inches) apart requires 250,000 seedlings. A backbreaking task done in muddy water, it takes four people more than six days.[9] People often work in teams; in some villages, families exchange labor.

In Asia, a farm population of more than one billion depends on rice as its mainstay.[10] By and large, each farm family works tiny plots that do not add up to much more than a hectare. In these plots, farmers produce almost half a billion tons of rice, enough to feed half the planet.

Diversity

Rice has been produced longer, by more farmers, and across a greater range of environments than any other crop.[11] Archaeological remains suggest it was cultivated in Asia nine thousand years ago.[12] Where did the crop originate? Probably along the great rivers of South and Southeast Asia: the Brahmaputra of northern India, the Irrawaddy of Burma, the Mekong of Indochina, and the Yangtze of China, all of which begin in the Himalayas. This great rice corridor stretches from southern China to the Ganges. Few regions on the planet are more diverse culturally; no region is more diverse in rice environments.

In areas subject to flooding, farmers in Indochina plant deepwater rices. They can survive submergence for up to two weeks without ill effects; many grow fast enough to keep their heads above water. Remarkable in this respect are the "floating" rices of Burma and Bangladesh, which can elongate more than 100 millimeters a day, reaching lengths of up to 6 meters. If the waters do not recede, harvesting is done by boat. For irrigated environments with controlled water levels, farmers plant paddy rices. There are special rices geared to rain-fed conditions. Rice can even be grown as an "upland" crop in dry fields, but distinct upland varieties must be used. In hillside fields, for example, farmers broadcast upland rice seed directly into dry, prepared fields, just as for maize, sorghum, or beans. In general, a wet rice type will do poorly in upland fields and a dry or upland rice will not flourish in standing water.[13]

The climatic range of rice is likewise impressive. Rice grows at sea level in Bangladesh and at altitudes of three thousand meters in Nepal. Rice flowers in the intense heat and humidity of the Ganges Delta, in the semiarid tropics of southern India, and in temperate highland valleys. During the rainy season, rice grows under a dense cloud cover, and in the dry season under high solar radiation.[14]

For a hundred centuries farmers selected seed from the rice plants best suited to their local soils, topography, climate, and rainfall pattern. They selected for shape, size, color, and taste. Over time, the great divide in the rice world of *Oryza sativa* emerged: that between indica and japonica plant types.[15] Thin and elongated, the indicas spread south into tropical climates. When cooked, indica rice is usually dry, firm, and fluffy. That is because of its high, starchy amylose content. Japonica types, by contrast, spread north from the Yangtze Valley. Along the way, they became more tolerant of cold weather than the tropical indicas. Japonica grains are short and stubby with a low amylose content; when cooked, the rice is stickier than indica types and tends to be mushy. India and Southeast Asia prefer indica types; Japan and Korea favor the japonicas. In China, the Yangtze Valley is the dividing point: japonicas to the north and indicas to the south. Americans usually buy intermediate types closer to the japonica side of the rice divide. For a true indica, such as aromatic basmati rice from India, or a "sweet" japonica from Koda Farms of California, they go to a specialty store rather than the supermarket.[16]

In warm temperate regions, most local rices are photoperiod sensitive: they flower and ripen as the days start to get shorter. Consequently, the maturation date is fixed. No matter how early farmers plant, the rice will not flower until the days are short enough. Closer to the equator, where day length differs less, the plant's habits tend to be photoperiod nonsensitive: neutral with respect to day length. The life cycle of such rices is independent of day length. Over the course of the growing season, some Asian farmers plant both types.[17]

The fit between a specific environment, a type of rice, and a local farming practice creates a complex picture. Farmers in Bangladesh, for example, plant at least seven different indica rice types. For the irrigated winter boro crop, planted in December and harvested in May, they grow cold-tolerant indicas transplanted from seedbeds. With the irrigated summer Aus crop (April to August), farmers either broadcast the rice seed directly into dry fields before the monsoon rains begin or transplant into flooded paddies. In rain-fed paddies, farmers plant an autumn aman crop from July to September; they use photoperiod sensitive varieties, which they usually transplant. Traditional deepwater rices, broadcast directly into dry fields, can take up to nine months to mature. Farmers seed fields in March or April, depending on when flooding occurs, and harvest them as the waters recede between September and December. Where flooding is deep, from one to six meters, and sustained, from three to six months, farmers need floating varieties.[18]

To preserve the crop's genetic inheritance, IRRI collects evaluates, and stores thousands of rice varieties every year. It is an enormous undertaking. It is estimated that there are about 100,000 varieties of *Oryza sativa*.[19] In 1992 IRRI's "germ plasm bank" had 74,500 *Oryza sativa* entries from 113 countries.[20]

Diffusion

Not all rices are Asian. Africa, Europe, and the Americas have local, heirloom varieties (or landraces) their farmers selected. In fact, West African farmers domesticated a different species around 1500 B.C., the savanna rice, *Oryza glaberrima*. Cultivation probably began in the central flood plain of the Niger River. Agriculture depended on the river's seasonal flooding, which created vast marshes ideal for diversified production. Farmers planted many glaberrima rice types, including deepwater, floating, and upland varieties. From the Niger Valley, rice culture spread to the Atlantic marshlands of West Africa between Senegal and Sierra Leone. In the sixteenth century, the coastal region had extensive irrigation works for transplanted rice.

Portuguese spice traders probably brought *Oryza sativa* indica rices from Asia to West Africa. The new rices spread quickly, often mixed with local favorites. To this day, farmers still plant both glaberrima and sativa rices, although varieties of Asian ancestry now predominate. In Nigeria and the Ivory Coast, currently West Africa's leading rice producing countries, farmers cultivate mostly unimproved *Oryza sativa* cultivars.[21]

Rice has a long history, even outside its Asian-African homeland. During the Middle Ages, it was an important crop in the Mediterranean world, including Italy, Spain, and the Levant. In the sixteenth century, Italy's Lombard landlords expanded production by draining valley swamps and grading paddies. Today, Italy's temperate Po Valley is Europe's leading rice zone, favoring cold-tolerant japonica types.[22]

In 1750, rice was grown in the New World, from Portuguese Brazil to Spanish Peru and English South Carolina. Brazilian production was in the tropical wetlands of Guanabara Bay near Rio de Janeiro and in the Amazon basin of Pará. In the 1790s, Brazil exported considerable quantities of rice to Portugal; in 1796 alone, some 176,000 arrobas or about 2,000 metric tons.[23] Brazil is currently the top rice-growing country in the Americas, but its production has moved south to states with more temperate climates, particularly Rio Grande do Sul.

The Moors brought rice to Spain. In the sixteenth century, Valencia was the country's principal rice-producing area. From Spain, rice made its way to the New World. It was an important colonial crop along Peru's Pacific coast near Trujillo, but for domestic consumption rather than export. In Spanish America today, Colombia produces the most rice, twice as much as its closest rivals, Peru and Ecuador.[24]

In the United States, rice was first produced along the rivers and swampy inlets of coastal South Carolina. Introduced in the 1690s, the crop's rapid spread depended on slaves from rice-growing West Africa who knew how to cultivate, process, and store the crop. Irrigation systems along the swampy, West African littoral were comparable to those built in colonial South Carolina. The use of rice mortars for dehulling, plus the design of fanner and storage baskets likewise suggest a West African connection. Subsequently, rice production advanced south into Georgia.[25] These states kept the lead in rice until the 1880s. Thereafter, Louisiana become a major competitor, producing more rice than any other state between 1890 and 1950. Since then, rice has moved up the Mississippi Valley as far north as Missouri. Arkansas currently leads the United States in rice production, followed by California's Sacramento Valley.[26]

Even Australia grows rice, although until recently not very much. In 1914, a Japanese settler brought seed to the basin along the confluence of the

Murray and Darling Rivers beyond Adelaide. In the 1950s, rice production was only about 100,000 metric tons. In 1994, however, it exceeded 1 million tons.[27]

How farmers produce rice depends on the rice environment, the local culture, and farm size. No crop in the world is grown so successfully in so many different ways. Beyond the coastal swamps of West Africa, rice is mostly an upland crop produced by small farmers. They mix rows of rice with sorghum and cassava (maniac) as part of a multicrop, subsistence farming system. In Brazil, by contrast, small farmers produce irrigated rice as a cash crop. Unlike in Asia, Brazilian farmers do not transplant rice. Instead, they broadcast their rice seed into soggy fields. As for upland rice, it is a crop for big ranchers in Brazil's interior.[28]

In Asia, most farmers transplant rice to irrigated fields, but their plots are small. In Japan, the typical rice farm is about 1 hectare in size; in Indonesia and India, it is often less. In the United States, the average size of a rice farm is 150 hectares; the crop is always irrigated but never transplanted. Farmers in south Louisiana ratoon their rice for successive, though smaller, harvests. In Arkansas, rice is drill seeded in a dry bed. In California, pregerminated seed is broadcast by aircraft over shallow standing water. Given its dry climate, high solar radiation, and low disease buildup, California's yields are 9.4 tons per hectare, among the highest in the world.[29]

In Asia and Latin America, almost all the rice produced is eaten as a grain. In the United States, 20 percent ends up in beer and another 21 percent in processed food. The largest rice consumer in the United States is the beer producer Anheuser-Busch, and Michelob, Budweiser, and Coors all add rice to their barley malt.[30]

Despite production elsewhere, no region of the world is as dependent on rice as Asia. In 1994, it accounted for 91 percent of the world's rice production.[31] Monsoon Asia has been the most densely populated place on the planet for thousands of years. It is not surprising, therefore, that a report on rice is first and foremost a report on Asia's rice belt.

KEEPING AHEAD

Between 1950 and 1990, the world's population growth exceeded 2.7 billion. Almost two-thirds of the increase was concentrated in Asia, particularly in the monsoon rice belt.[32] Keeping food production ahead of population growth is a great achievement. For the 1970s, many experts forecast widespread famine in Asia.[33] And with good reason. Food production had stagnated. Asia's rice harvest averaged 203 million tons in 1956–57; five years later, in 1961–62, it had increased to just 204 million. Compared to the region's rapidly growing population, the rice harvest was standing still. Specific countries, of course, did better than the average. Between 1956 and 1960, India's rice production averaged 45.5 million tons; for the years 1961–65, harvests were up significantly, by an average of 14 percent. Nonetheless, India's population grew faster, from 394 million people in 1956 to 486 million in 1965—an increase of 23 percent. For Indonesia in the same period, harvests went up an average of 6 percent; the population, however, grew from 86 million to 104 million, more than three times as fast. Meanwhile, in China, with the world's largest population, production hardly changed. For 1956–60, harvests averaged 72.2 million metric tons; for 1961–65, 75 million tons—an increase of just 4 percent. Given statistics like this, it is not difficult to see why the experts predicted famine.[34]

Fortunately, the outcome for rice did not follow the gloomy trends forecast. In 1960, Asia harvested 202 million metric tons of rice. Ten years later, the total had jumped to 290 million tons; in 1980, it was 362 million, and in 1990, 479 million—an overall increase of 137 percent, which for the moment put Asia's food production a big step ahead of population growth. The trend applied to Asia's largest rice producers. India's rice crop more than doubled, from 52 million metric tons in 1960 to 110 million in 1990. Indonesia's production tripled, from 12.8 to 43.8 million tons. And in China, which irrigated almost all its rice, harvests increased from 80 million metric tons in the mid 1950s to 188 million in 1990.[35]

Modern Varieties

What made the gains in Asia's rice production possible? The key factor is the creation of "modern" semidwarf rice varieties. Japan developed the first high-yielding, modern japonicas, including the ponlai types, which it introduced into Taiwan during the 1920s. These were shorter than traditional rices, stiff-stemmed, and responded to fertilizer by setting more grain. Japonica types, however, do not do well in the tropics. They flower too early, put out fewer stems, and are highly susceptible to tropical diseases.

Traditional indicas rices are between 1.5 and 2 meters tall with long, weak stems. When nitrogen fertilizer is added, the plant sets more grain than the stems can support. The heavy grain heads make the plant fall over or lodge. The monsoon rains and wind further beat the plants down into the mud. To breed a better rice type for the tropics, the United Nations Food and Agricultural Organization (FAO) sponsored an indica-japonica hybridization program in the 1950s at India's Central Rice Research Institute. The objective was to cross the japonica dwarfing gene into an indica plant without sacrificing the traits needed in tropical environments. Results were not satisfactory, at least compared to IR8.[36]

The first modern variety IRRI released in Asia was IR8. It was developed by Peter R. Jennings, Te-Tzu Chang, and Henry M. Beachell. Jennings was a plant pathologist and breeder trained at Purdue University. Before heading IRRI's breeding program, he worked with the Rockefeller Foundation as a rice specialist in Colombia. Born in China, Te-Tzu Chang was trained in agriculture at Cornell University and the University of Minnesota. He was a senior agronomist with the Rural Reconstruction Commission in Taiwan when hired by IRRI in 1961. Beachell was a rice breeder at the Texas research station in Beaumont, where he had trained Jennings for his assignment in Colombia. Together they worked out the breeding program that led to IR8. They wanted short, sturdy-stemmed indica varieties that did not fall over, even with high levels of fertilizer application. This time, the dwarfing gene came from a mutant indica rice, Dee-geo-woo-gen, which Chang brought from Taiwan. Using the bulk breeding approach, semidwarf indicas were crossed with tall ones. The result was IR8. According to Robert Chandler, IRRI's founder and first director, "It was Chang who pointed out the importance of the short-statured indicas from Taiwan, Jennings who selected IR8's parents and made the cross, and Beachell whose keen eye picked out IR8 from a multitude of segregating lines."[37]

Tested in 1965 at various sites in the Philippines, Thailand, Malaysia, and Taiwan, IR8 was released in 1966, initiating what was soon called a "green revolution" in Asian rice production. Chinese scientists, meanwhile, had developed semidwarf rices independently, even before IRRI. In 1965, the new varieties were already on more than 4 million hectares in southern China. The results of China's rice research, however, were not made public; IRRI's were.[38]

What made IR8 so much better than traditional indicas? It was much shorter, less than a meter tall, with thick, sturdy stems that held the plant erect. It responded to nitrogen by putting out more tillers (extra stems), which produced more seed-bearing panicles, which in turn set more grain—in fact, twice as much. And the strong stems did not lodge, not even in the wind and rain. It matured more rapidly, in about 125 days, and it was insensitive to daylight. But IR8 also had some disadvantages. It had a chalky, coarse grain, it was sticky, and it tended to harden when cooled, characteristics consumers did not like. Moreover, IR8 was not sufficiently resistant to important diseases and insect pests. So it needed improvement.[39] In 1969, IRRI released IR20, which had better grain quality and more pest resistance.[40] Advantageous traits were continuously added. IR36, released in 1976, had resistance to some fifteen insects, diseases, and environmental stresses. It had long, slender, translucent grains and matured in just 110 days. In 1982, IR36 was grown on 11 million hectares of rice land under a dozen different names, probably making it the most widely grown crop variety anywhere in the world.[41]

Rice breeding today involves IRRI and its many partners in Asia. Whereas IR8 had just three parental lines in its genetic background, IR36 had genes from thirteen different varieties and a wild species, *Oryza nivara*. Released by the Philippine Seed Board in 1988, IR72 had eighty-seven parents from eight countries in its lineage.[42]

Soon IR8 was replaced, but its demonstration effect was unequaled. No one thought indica rices could yield more than 3 tons. Suddenly, there was a rice plant that produced 5 to 6 tons in the cloudy rainy season and 7 to 9 tons under highly favorable conditions in the dry season. That changed how farmers, scientists, and government officials thought about rice. Food security in Asia now seemed possible.

Farmers led the way. In 1966, to promote IR8 in the Philippines, IRRI gave away 2 kilos of seed to any farmer who came to the institute and picked it up. According to Chandler's account, in six months, more than two thousand farmers came to IRRI—by bus, bicycle, and on foot. Eventually IR8 seed was planted in forty-eight of the country's fifty-six provinces. The new seeds spread quickly because farmers wanted them.[43]

To do their best, short-statured rices like IR8 need carefully managed water levels. Yields twice, even three times, greater than before justified expanding the acreage irrigated, especially where it was cheap and easy to do so. Between 1960 and 1986, India increased its irrigated rice acreage by almost a third, from 12.4 to 18.1 million hectares. In the Philippines, the area doubled, from 960,000 hectares in 1960 to more than 2 million in 1989. For Indonesia, it rose from 5.7 million hectares in 1972 to more than 8 million in 1987. In 1990, 55 percent of Asia's harvested area was irrigated, a favorable environment that accounted for three-fourths of Asia's rice production.[44]

The payoff for using modern varieties is greatest with irrigation—but not exclusively so. Some rain-fed zones have optimal conditions. The monsoons bring enough rain for diked fields to catch and hold the water. When the water level is high enough, the rice is transplanted into the paddy. Nonetheless, where the monsoons are not dependable and droughts strike, or where flooding occurs, modern varieties offer much less to farmers.

Averaging across the diverse environments in which rice is produced, the results are still significant. Between 1960 and 1990, Asia's yields per hectare doubled from 1.8 to 3.6 metric tons. Total production rose from about 200 million to almost 480 million metric tons. That was much greater than the increased amount of land Asia devoted to rice during the same period, which expanded by only a quarter.[45]

The relationship between a country's yields, modern semidwarfs, and irrigation is straightforward. Farmers in China and Japan, for example, irrigate more than 90 percent of their rice fields. For 1992–94, they had some of the highest average rice yields in the world: 5.8 tons and 5.9 metric tons per hectare, respectively. Both countries rely almost exclusively on modern varieties, most of which they pioneered on their own. Indonesia irrigates about 80 percent of its rice crop; farmers use modern varieties on 85 percent of the country's rice land, and overall yields for 1992–94 averaged 4.4 tons. By contrast, India's yields per hectare averaged only 2.7 tons. But India irrigates just half its rice area, and only 58 percent is planted to modern varieties.[46] Nonetheless, the gain over 1962–64, when yields averaged just 1.5 tons per hectare, is considerable. An extra ton averaged over the 42 million hectares India harvested in 1994 adds up to a lot of rice.[47] And India has made even greater gains in wheat.

It costs farmers more to plant modern varieties. To get high yields, they apply nitrogen fertilizer in quantity. In India, usage went up from an average of 2 kilos per hectare in 1960 to more than 30 kilos in 1979. During the same period, Indonesia increased its application from an average of 8 kilos to more than 57 kilos.[48] For most Asian countries in 1990, on irrigated rice fields, farmers typically applied 60 to 90 kilos of nitrogen fertilizer per hectare for the rainy-season crop and 100 to 150 kilos for the dry-season crop. The investment pays off in yields, particularly when the cost of fertilizer is subsidized.[49]

Even without fertilizer, modern varieties outyield traditional ones. Breeders keep adding other advantages too. Modern varieties do better under climatic stress and resist more diseases and insect pests. They also mature more rapidly, some in less than 100 days; traditional irrigated indicas need upwards of 150 days to reach maturity. Rice production has risen not only because yields are high but also because early maturity gives farmers time to fit in two, even three, rice crops a year. That is a fourfold and sometimes even a sixfold increment in total rice production. Semidwarf indicas are typically photoperiod insensitive, so they can be used at latitudes where seasonal day length changes significantly, as in northern India, Bangladesh, and most of China.[50]

Getting High Yields

The rice revolution took place on many fronts; IR8 is only part of the story. To get the most from high yield potential, farmers have to be fussy about preparing their fields, spacing plants, weeding, water control, and fertilizer use. New varieties grown without the recommended technology produce much less. To have an impact, the rice revolution had to change how millions of farmers in the paddies of Asia worked, and that required local extension and government support. What the breeders did was only part of the picture.

In the 1970s, Robert Chandler analyzed which factors contributed the most to high yields in developing countries. As might be expected, the use of irrigation, semidwarf varieties, and new production technology were high on his list. His success stories then were Taiwan and South Korea. In 1973, Taiwan had 540,000 hectares in rice, 98 percent of it irrigated. The modern variety then preferred was Tainan 5, a short-statured japonica type. It had short maturity, from 95 to 120 days, which allowed Taiwanese farmers to plant two rice crops a year. In fact, 60 percent of the country's rice fields were double-cropped. Between 1952 and 1975, yields increased from 2.5 to 4.5 tons per hectare. In the meantime, fertilizer use doubled to 114 kilos of nitrogen per hectare. Farmers spent $33 million on pesticides in 1975, compared to almost nothing in 1952.[51] Overall rice production increased from 1.9 million metric tons in 1951 to 3.5 million in 1976.[52]

Behind the success story, however, were factors independent of production technology. In the mid-1950s, the government had undertaken a land-reform program that transferred landownership to tenants. To make fragmented, small holdings productive, it consolidated plots for joint farming. Moreover, Taiwan had invested heavily in its own rice research program. Tainan 5, for example, was developed in Taiwan. Agricultural was considered central to the country's development strategy. The government subsidized prices for fertilizer, pesticides, and farm machinery and strengthened local farmer associations. In 1976, Taiwan had 273 farmer-managed township associations subdivided into thousands of village units. The associations employed more than sixteen hundred extension agents, provided more than $350 million dollars in credit to farmers annually, and had accumulated $470 million dollars in deposits. They owned four hundred rice mills and sixteen hundred warehouses with a total storage capacity of 850,000 tons of rice. The associations sold seed and agricultural supplies, including 400,000 tons of fertilizer. Finally, Taiwan had sixteen self-governing, farmer-led irrigation associations, which managed 21,000 miles of canals and ditches. Chandler considered Taiwan's irrigation system the best operated and maintained in Asia.[53]

South Korea had worked on land reform and rice production since the mid-1950s. Between 1966 and 1968, it spent more than $60 million dollars on irrigation projects; in 1970, 85 percent of its rice acreage was irrigated and it had high yields of 4.6 tons per hectare. South Korea relied on modern varieties, most of them japonica types adapted to the country's temperate climate. Nonetheless, it was still dependent on rice imports, which in 1971 exceeded a million tons.[54]

Between 1972 and 1976, South Korea increased its average yields by 1 ton per hectare. In part, success resulted from a joint breeding effort with IRRI. Korea's japonica varieties had poor disease resistance and a low threshold for nitrogen: high quantities of it made them lodge. An improved japonica-indica cross was released to farmers in 1972. Just as important, however, was South Korea's self-help, village-level development program launched in 1970. Under the aegis of the New Village movement, drainage ditches were improved, wells were dug, roads were built, and warehouses were constructed. The program provided materials and supplies, villages contributed their labor. Between 1972 and 1977, the number of cooperative farming units increased from twenty-two to fifty-two thousand. Backed by the cooperatives, farmers worked more efficiently, sharing their equipment and labor. At the same time, the ranks of the extension service expanded to meet the new demand for technical assistance: from 1,800 specialists to 7,500. In 1976, the extension service gave short training courses attended by some 2.8 million Korean farmers. For the years 1977 through 1979, South Korea's rice yields averaged 6.7 tons, the highest in the world. And the country was virtually self-sufficient in rice production.[55]

For high yields and food security in Asia, adopting modern varieties was just a first step. The rice revolution is also about credit programs, technical assistance, and farmer-led organizations.

Impact

When IRRI was founded in 1960, production meant everything. More rice quickly was the objective. It focused on an improved indica rice type for tropical Asia's irrigated, transplanted fields. The impact IR8 and its successors had on production was dramatic. But how did the benefits get distributed between rural areas and the cities, between favored areas with irrigation and unfavored areas with uncertain rainfall, between landlords and sharecroppers, between early and late adopters? Once the food crisis had passed—at least temporarily—the thorny issue of equity arose. Researchers wanted to know who adopted the new varieties first and who benefited most?

A 1985 World Bank impact study concluded that semidwarf yields in Asia were so much higher that farmers converted rapidly, whether they were small landholders or large producers, and whether they owned their own land, rented land, or were sharecroppers. Modern varieties meant more food at lower prices, a significant gain for the poorest third of Asia's population, which spent 60 to 80 percent of household income on food. Since IR8 took hold, the price of rice on world markets has fallen 40 percent and per capita rice consumption has gone up 25 percent.[56] For most rice farmers, the yield increase more than compensated for the drop in prices, so they benefited as well. And in many cases, farmers used fertilizer and pesticides whose cost was subsidized. As for wage workers in the fields, growing modern rices required more inputs, and often led to double-cropping, which thereby increased the demand for labor.[57]

Such is the general picture. Examined locally, the conclusions sketched above have to be modified. For monsoon Asia is diverse in its rice environments, in its cultural beliefs, and in the way the countryside is organized. A study of the North Arcot District in the Indian state of Tamil Nadu compared the early 1960s with the mid-1980s. It found that the demand for labor in rice cultivation had fallen 4 percent—despite the increase in the district's irrigated rice paddies. Agricultural earnings, however, had doubled, even for landless laborers. And the multiplier effect on the district's nonfarm economy was substantial, due to increased farmer demand for production inputs, marketing services, and food processing and because of the extra consumption that resulted from higher incomes. Land concentration did not change significantly during the period. Small farms of about 1 hectare in size still predominated. And the income inequality within the farm sector did not change.[58]

North Arcot, of course, is a single case. How typical is it with respect to India overall? A 1988–89 survey reached somewhat different conclusions. Despite India's undisputed gains in rice production, yields and, in some states, employment, the green revolution did not reduce rural poverty significantly. Nonetheless, there was little evidence of either reduced wages or a drop in the demand for labor. Moreover, the new rice technology helped make small holdings viable, stemming migration from the villages to the cities.[59]

For Asia as a whole, recent studies have reexamined the impact that modern rice varieties had across countries. The main equity distortion they found was a widening productivity gap between environments whose conditions favor the use of modern varieties and environments whose conditions do not. Consequently, farmers in irrigated areas had gained more than farmers in risky rain-fed areas or those in upland zones. How great is the gap in per capita household income between rice-producing environments? In the Philippines the gap is wide; in China it is remarkably small. For India, differences are intermediate overall, with considerable variation by state. In Indonesia the income differential is small, in part because farms in rain-fed areas are larger.[60]

The Future

The experts of the 1960s were wrong about food production, but they were correct about the world's population growth. Between 1950 and 1990, it went from 2.5 to 5.3 billion people.[61] Why was it that countries successfully transformed rice production but failed to reduce population growth? The transition from high to low fertility depends on religious factors, economic conditions, and social customs. A decline in the size of families occurs at different rates in different places for different reasons. In China national policy permits only one child per family. In India, state governments support health clinics and family planning, but they rarely impose family size by decree. In any case, changes in material culture are not always good predictors of changes in what people think. Improved rice production technology can be adopted with little impact on fertility behavior.[62]

In fact, between 1965 and 1990, population growth rates dropped in many poorer countries, including Asia's rice belt. South Korea, Taiwan, and Singapore, for example, all high-fertility countries in the 1960s, have rates of population growth today comparable to Canada. In Brazil, Colombia, and Sri Lanka, the population growth rate dropped by almost half. For India, Burma, and Bangladesh, the rate also fell, but not as dramatically. In global terms, the world's population was growing at a rate of 2.2 percent in 1965 but had dropped to 1.7 percent in 1990, a 20 percent decline. Nonetheless, most poorer countries have very young populations. Even if couples have fewer children than their parents did, the population will continue to grow. The planet will have six billion people by the year 2000 and perhaps eight billion by 2020.[63]

Almost half that population increase will be in the rice belt. The region today is self-sufficient in rice, but can it feed an additional 1.5 billion people? For Asia overall, that means an increase of more than 50 percent from the 485 million metric tons of rice produced in 1994 to more than 750 million tons for 2020.[64] Is such an increase possible? Can it be done in a sustainable way without undermining future productivity? Maybe.

Because not every tack is likely to pay off, IRRI approaches rice production with complementary strategies. Pest control is an example. One solution is to breed a resistant plant. Another solution is pest management: limiting insect damage biologically by building up the population of predators and pathogens. Whether it is higher yields, weed control, or pest management, IRRI works on two broad fronts: breeding a better rice plant and improving how rice is produced. To better understand IRRI's ongoing research, these two fronts must be kept in mind. The section that follows explains why particular breeding objectives are important and looks at ways to increase on-farm yields. It is divided into three parts: genetic improvement, production technology, and on-farm research.[65]

GENETIC IMPROVEMENT

The key advantage modern rices had over traditional ones was in yield, a factor that held up throughout Asia's irrigated rice belt. However, once a country had built up its rice stocks, grain quality became more important.

Rice preferences vary from country to country, even village to village.[66] For consumers, how the rice cooks, whether it ends up soft or firm, sticky or fluffy, is high on the preference list. For noodles, a firm rice is desirable; for puddings and porridge, a soft, mushy one. Farmers who rotate from rice to other crops want a rapidly maturing rice. Because diseases, pest problems, soils, and climatic conditions can differ from one locality to the next, a variety that is the average best fit for the rice belt overall is not the best for every location. Consequently, to benefit from IRRI's work, Asian countries need their own rice research programs. At national centers, scientists work out production technology for difficult climates and cropping patterns, and they isolate traits in local cultivars that farmers and consumers want in modern rices. According to IRRI scientists, a microenvironment that covers at least 100,000 hectares of rice merits its own breeding program. Such specialized work requires strong, local research.

Networks

How can IRRI reconcile a broad breeding focus with the narrower objectives a national program has? The answer is to make testing new varieties a joint venture. The first research network IRRI established, the International Rice Testing Program, evaluated advanced rice lines at sites all over Asia.[67] Today, testing is coordinated by the International Network for Genetic Evaluation of Rice (INGER).

What does the INGER network accomplish? Consider genetic resistance to a pest like plant hoppers. Scientists have to expose each new rice line to different levels of infestation and to several of the insect's biotypes. Working alone, it can take decades to go through the many combinations possible. With a network, local breeders can test the lines against a range of pest levels and biotypes at rice production sites across Asia.[68]

In 1990, the plant hopper nursery included about 150 entries or "lines" selected by the network's coordinating committee. All the participants, including breeders at IRRI, agreed to plant out the full nursery, expose it to plant hoppers, and record the results, which are subsequently published and distributed. In this way, a local breeder can get his rice line evaluated in enough localities to save years of work. Even if his own entry does poorly, he can select new lines from the varieties that do well. Countries are free to release any rice variety from the trials, under any name.

There are more than twenty distinct testing nurseries under IRRI. New rice lines can be screened for disease and pest resistance, for tolerance to abiotic stresses such as saline soils, and for adaptability to different rice environments. Dr. Durvasula Seshu, a tall, dedicated man in his fifties, coordinated each step.[69] To enter a line, countries send a 100-gram (100 grams = 3.5 ounces) packet of seeds to IRRI for multiplication. Because some 30 to 40 kilos of seed per entry are required, IRRI multiplies seed for every entry in every nursery every year. And it has to be high-quality seed with a good germination rate. Then the seed has to be distributed so it reaches testing sites at the right time.[70] According to Seshu, that means tracking the growing seasons in rice-producing countries all over the world.

In 1992, 1,300 entries from 19 nurseries were distributed to 35 countries in Asia, sub-Saharan Africa, and Latin America. Overall, since 1972, some

International Network for Genetic Evaluation of Rice (INGER)

STRESSES

DISEASES
- Blast
- Bacterial blight
- Tungro virus

INSECTS
- Brown plant hopper
- Whitebacked plant hopper
- Stemborer

ADVERSE SOILS
- Salinity
- Alkalinity
- Iron toxicity

TEMPERATURE
- Heat tolerance
- Cold tolerance
- Heat susceptibility

MOISTURE
- Submergence
- Drought at tillering
- Drought at flowering

RICE ENVIRONMENTS

- Irrigated lowland
- Rain-fed lowland
- Rain-fed upland
- Floating rice
- Deepwater rice
- Swamp rice

14,000 entries have been tested at more than 700 locations in 40 countries.[71] Collecting, multiplying, and distributing seed is IRRI's job, a cost covered by a grant from the United Nations Development Program (UNDP). Participating countries plant, harvest, and evaluate the nurseries they request. The network cuts back on the time needed to breed new lines with the right traits. As of 1992, the network had generated 175 new varieties, which came from 18 different countries, and which had spread to more than 50 nations.[72]

Biotechnology

In rice breeding, isolating a complex trait such as salinity tolerance is just a start. The trait has to be crossed back to a high-yielding variety and then tested at different trouble spots. If the tolerance holds up, millions of kilos of seed must be multiplied for distribution to farmers. Because it can take a decade to go through all the steps, a variety being released today is often a response to priorities already out of date. For complex traits, biotechnology makes conventional breeding faster and more efficient.

For a few rice traits, breeders have morphological markers. The gene for resistance to brown plant hoppers is a case in point. In some varieties, it is closely linked to a gene that gives a purplish tinge to the rice plant's leaf shoots. When breeders make a cross, they can select for plant-hopper resistance by checking for shoot color. This eliminates the arduous task of infesting thousands of rice plants with plant hoppers. Unfortunately, morphological markers, especially for complex traits, are rare.[73]

Dr. Susan McCouch, a molecular biologist trained at Cornell, tags rice chromosomes using Restriction Fragment Length Polymorphism (RFLP) probes.[74] She was at IRRI with her husband and two small children. In McCouch's research, identifying an RFLP marker is analogous to finding a morphological one. The lengths of the restriction fragments at a particular site on the rice genome differ between varieties due to mutation. To measure a fragment's length, DNA is first extracted from a plant, transferred to a test tube, and then cut by enzymes. As McCouch explained, the objective is to link RFLP markers to a variety's economically important traits. When a cross is made, if a marker cosegregates tightly with a trait, the trait or gene is said to be tagged.[75] "When we know which RFLP markers are close to a specific gene, then we can select for the trait by selecting for the RFLP markers," McCouch said.

With DNA-based genetic mapping, researchers may one day check for rice traits by looking at the markers on strands of chromosomes. Consider screening for drought tolerance. Researchers use special equipment to measure root depth and concentration, an assessment made at different points in the plant's life cycle. Much time could be saved if scientists knew which genes controlled the trait; they could then screen for drought tolerance in the lab, analyzing a plant's DNA with RFLP probes. Breeders would still need field trials to make sure they were right, however, for genetic inheritance is a probability, not a certainty.

"An RFLP marker is an extension of our vision," McCouch said. "A picture on a microscopic level." Before such pictures can be used routinely, the genetic sequence they portray will have to be much more exact. That will take years of work. Each cross made requires an independent mapping study. Consequently, the parents crossed must be chosen carefully. To make the study worthwhile, the traits of interest must be difficult, time-consuming, or expensive to screen for directly. Major RFLP mapping programs are underway at public institutions in four countries; gene tagging studies are done at numerous locations worldwide. IRRI brings the results together, which it shares freely with rice scientists everywhere.[76] So far, using cosegregational analysis, more than nine hundred RFLP markers have been placed on the genetic map of rice. In addition, more than twenty single genes for disease and insect resistance are now located relative to RFLP markers.[77] When a map of the rice genome is finally put together, it will be easier to assemble traits that make for the right plant in the right place. Some scientists believe the technology will help diversity the genetic makeup of the rice cultivars in use.[78]

Research using RFLP is very expensive. It can make conventional breeding more efficient, but it is no substitute for it. In the meantime, durable, if less glamorous, techniques are in use. Embryo rescue makes it possible to cross rice with wild relatives, species that retain valuable traits lost in domestication. Salinity tolerance, for example, was recently crossed into rice from a distant grassy relative adapted to saline soils. Darshan Brar, a soft-spoken, turbaned scientist from India, works on such wide crosses at IRRI's biotechnology unit. He explained that in the field a cross between wild and domestic rice aborts.[79] To rescue the embryo, it is surgically removed ten days after pollination and put in a test tube with a medium for nutrition. Once the embryo germinates, they strengthen the plantlet and then move it to the field. "This sounds easy," he said, "but sometimes we have to try thousands of times, year after year." It is worth it. The technique has accomplished much already, including a plant with salinity tolerance. To start, researchers needed just one tolerant hybrid plant. Then they backcrossed it to a cultivated variety for pollination. This cross gave them hundreds of seeds, which in turn generated a population of new plants. From these they selected those that retained the trait for salinity tolerance. They rejected the rest, ninety-nine out of one hundred.

A plant that tolerates saline soils has much to offer. Even when rice is irrigated, saline soils keep down yields to about 1 ton per hectare—far below the potential of modern varieties. South and Southeast Asia alone have more than 5 million hectares of such soils. For rice, salt water intrusion compounds the problem, particularly in Asia's tidal wetlands. In Bangladesh and Indonesia, the swampy lands along river deltas are unsuited to virtually every crop except rice. When the monsoons bring enough rain, the heavy fresh water pushes back the denser salt water. But when the rains are less intense, salt water can move up the many small streams and river channels that drain the swamps, doing great damage to the rice crop.

According to Dr. Dharmawansa Senadhira, breeding for such an environment is difficult "because the degree of salinity, both in the soil and water, changes from place to place." Senadhira, born in Sri Lanka, works on breeding for environments in which modern varieties have not made much headway.[80] He is a patient man, and necessarily so, for salinity tolerance is tricky and frustrating. Not only do the salt concentrations in the water and soil change during the rice-growing season, but saline marshlands are subject to tides that extend up inland channels for as much as 70 miles. The traditional varieties suited to this condition are very tall and low yielding.

In tidal wetland environments, salinity tolerance is just one of the traits important to farmers. So breeding is a joint effort with many countries, whose breeders make their selections locally. Exchange and testing new lines is done through the Saline Soils Nursery. By using anther culture, breeders can generate pure lines faster. A traditional hybrid cross is first done in the field. Then, in the lab, tiny anthers are taken from the rice floret and cultured on a suitable medium. The callus that forms can be used to generate homozygous plantlets and seeds. In this way, a cross can be stabilized in two generations.[81] When I talked with Senadhira, INGER had already found saline tolerant lines good enough to release. Farmers were trying them out at sites in Egypt and India.

Cold Tolerance

Before the semidwarf indicas swept across Asia's irrigated tropics, breeders in Japan and Taiwan had managed to get short-stature into cold-tolerant japonica rices. In many mountainous parts of Asia, however, indica rices susceptible to cold still predominate. At high elevations, India, Nepal, Bhutan, Bangladesh, and Indochina all face cold weather damage to the rice crop. Cold irrigation water from melting snow also retards the growth of sensitive indica rices. Even the Philippines and Indonesia, although much farther south, have cool, mountainous regions. Senadhira who also coordinated the cold-tolerance network, estimated that Asia's tropics had 7 million hectares of irrigated rice subject to cold spells.[82] In these areas, farmers still plant low-yielding indica varieties.

Cold tolerance is found in some indica rices, but the threshold is low. Japonica rices, by contrast, have a high threshold and do well even at altitudes of 2,500 meters. That being the case, why didn't farmers simply switch to japonica varieties? Because farmers and local consumers both preferred the less glutinous, thinner, long-grained indica rices. They just did not like japonica rice—the grains were too fat, too round, and too sticky.

One approach is to cross cold tolerance from a japonica rice to an indica plant, and from there, cross back to a high-yielding, modern indica. Unfortunately, the two rice groups are extremely difficult to cross. Most modern varieties are based on crosses within either the indicas or japonicas. Although some indicas have considerable cold tolerance, most are semiwild, low yielding, and hard to cross. Recently, however, a compatibility gene has facilitated hybridization. In conjunction with embryo rescuc and anther culture, it is possible to get cold tolerance from japonica rice types into some indica rices. "The challenge now," Senadhira said, "is to get the trait into a good plant."

That will not be easy, considering the different situations cold tolerance has to cover. In the high mountain valleys of Sumatra, in Indonesia, farmers want cold tolerance at the beginning of the growing season when they transplant. In eastern India (Bihar) and Bangladesh, cool weather can occur with the winter crop at midseason. In Nepal and Bhutan, farmers need cold tolerance throughout, from planting to harvesting. Every country has its own special requirements.

To evaluate new lines, IRRI screens every entry in a special greenhouse at temperatures that range from fifty-three to sixty-three degrees Fahrenheit. Nonetheless, evaluations had to be carried out locally, using INGER. Each country involved screened for cold tolerance with its own needs in mind. Thus, Indonesia screened at planting time, Bangladesh at midseason.

To screen for cold tolerance, the temperature of both the air and the water is monitored. Recall that rice environments are defined primarily by water regimen. Modern varieties spread first to irrigated fields where conditions were most favorable. They also did well in rain-fed zones, but fields had to be diked and dependable rainfall was needed to fill paddies to predictable levels. The rice revolution, however, bypassed swampy coastal wetlands and tidal basins; such environments were subject to saltwater intrusion or had saline soils. Even irrigation, which is generally favorable to rice production, presents exceptions, such as cold tolerance. And water—either too salty or too cold—is the key problem in both cases.

Deepwater Rice

Rice is so well adapted to water that it can grow in depths that would kill most other crops. The conditions that require deepwater, or "floating," rices are widespread enough to constitute a special environment. In East India, Bangladesh, and Burma, for example, there are more than 8 million hectares of deepwater rice, and in Southeast Asia, almost 3 million.[83]

Derk HilleRisLambers, a Dutchman who grew up in Indonesia, worked on the special problems that "flooded" rice poses.[84] In some cases, flooding is followed by a rapid drop in water levels. In others, the initial flooding is sudden but the water recedes gradually. Fast flooding is usually caused by heavy monsoon rains. "For this environment," HilleRisLambers said, "farmers need a rice that can hold its breath, that can tolerate submergence for at least ten days." The impact flooding has depends on how deep and how muddy the water gets. If the flood depth is not excessive and the water is clear enough for sunlight to filter through, the rice can survive a couple weeks without great damage.

In the typical deepwater pattern, rivers and tributaries rise gradually to spill over their banks and flood the lowlands. This occurs in delta areas, particularly in eastern India and Bangladesh, where the Ganges and Brahmaputra Rivers converge. The degree of flooding depends on rainfall farther upstream; eventually, flood waters gradually recede. For such conditions, farmers need tall deepwater rices with elongated stems that tolerate temporary submergence, varieties that can grow and stretch to keep their heads above water. Some floating rices of the deepwater type can grow 5 centimeters a day. The problem is that the plant's energy goes mostly to its stem rather than to seed production, so yields are low.

Some modern varieties are selected specifically for tolerance to temporary submergence, that is, to flash flooding with rapid water recession. Because periodic flooding is also common in rain-fed areas, the trait is advantageous over a wide region. Deepwater rices, by contrast, are characterized by the water depth in which they grow: 50 to 100 centimeters, more than 100 to 150, or more than 150 centimeters. They also take a very long time to mature—in excess of two hundred days. "For water levels above 50 centimeters," said HilleRisLambers, "modern varieties will not do; they are too short and mature too rapidly. So farmers stick with local varieties. Given the trade-off between height and productivity, the best we can do for now is a plant of intermediate height that has moderate yields and is adapted to depths of between 50 and 100 centimeters." Because local environments differ greatly, breeding is in conjunction with countries in which deepwater conditions are common. As is true of saline and cold tolerance, IRRI's INGER screens lines for tolerance to flooding and deep water.

Upland Rice

Not all rice is irrigated. With upland rice, farmers do not divide fields into paddies or flood them. They prepare and seed their fields the same way they plant maize or sorghum. Upland rice is produced in highland areas and on the plateaus, which get much less rainfall than the lowlands. Even in the lowlands it is grown on rolling land and hillsides not fit for irrigation.

Asia has more than 10 million hectares in upland rice.[85] Yields are often less than 1 ton per hectare, far below the 5 tons now common with irrigated paddies. So although farmers devote much acreage to the crop, the contribution to a country's total rice output is usually small. According to Dr. Michel Arraudeau, a hard-working, irascible Frenchman who breeds upland rice, although 10 percent of Indonesia's rice acreage, some 1.1 million hectares, is in upland varieties, upland rice accounts for only about 4 percent of the country's total production.[86] "The problem with most upland zones," Arraudeau noted, "is erratic rainfall with dry spells during the growing season. Soils are often infertile and highly acidic, and dry rice is particularly susceptible to the blast fungus. Compared to flooded paddies, weed problems with dry rice are much greater. So overall, dry rice is a low-yield crop."

Given the many adverse factors, from poor soils to weeds, why did IRRI bother with upland rice? Upland varieties are often specialty rices valued for their aroma, their cooking qualities, and their color. Farmers plant an astounding assortment: in Indonesia alone, more than two thousand varieties. So upland rice crops have an aesthetic appeal and help preserve genetic diversity.

Upland rice also has great ecological significance. Many farmers must plant their crops in rugged terrain prone to erosion. The depletion of the soil, the low yields, and the destruction of forests reinforce one another; as a result, upland farmers are usually a country's poorest. To make conservation viable and create incentives for crop management, better varieties with higher yields are needed. "What we want," said Arraudeau, "is a plant that tolerates drought better and withstands blast disease, a variety that is more productive in spite of acidic soils. If we can do this, we can double production." That is not much compared to yields from irrigated rice, but upland rice has to be looked at differently. What is most important is not the crop's yield, but the endangered ecosystem in which it is produced. The problem is not just how to grow more rice but how to work out rotations and cropping patterns that are sustainable, good for farmers and for the environment.

Germ-Plasm Banks

Breeders looked to IRRI's germ-plasm bank to find the rice traits they need. Dr. Te-Tzu Chang, who helped develop IR8, managed the bank. Born in China, he knows the history of rice and the crop's great diversity, which the germ-plasm bank tries to represent and preserve.[87]

To me, rice is rice; it all looks the same at the supermarket. Not so at the germ-plasm bank. The rices displayed range from purples, reds, and blacks to pinks, browns, creams, and assorted shades of white. They are long and thin, short and squat, round, oval, and flat. In 1993, the bank's total collection stood at 74,700 *Oryza sativa* accessions, 1,330 *Oryza glaberrima* varieties, and 2,216 wild species.[88] About 50,000 of these entries were catalogued and available for distribution at that time.

For each accession, the bank maintains "passport" data. The standard list contains some eighty items. It includes a variety's local name and country of origin, plus the latitude, longitude, and topography of the district where the rice originated. There is physical data: a plant's height, growth habits, and grain type, its vigor, days to maturity, and photoperiod sensitivity; there is resistance data: to assorted diseases and pests and to abiotic stresses, notably cold weather, saline soils, and drought; and there is production data: whether the rice variety collected is usually irrigated, grown under rain-fed conditions, or planted as an upland crop.[89]

For entry in the bank's collection, about 2 kilos of high-quality, disease-free seed is needed. Most of the time, IRRI has to multiply out a variety before depositing it. For Chang, producing a batch of seed for thousands of varieties is a messy, time-consuming business. There are typhoons every growing season, and they can do great damage. Great gusts of wind send the seed scattering. With the most delicate varieties, the panicles tend to split open or shatter. Getting enough high-quality seed for storage is a painstaking task. Selecting the best seed is a skilled job done by hand. Chang estimated that the Seed Unit could process about eighty varieties a week, depending on the quality of the seed received.

Seeds are dried to a specified moisture content and then packed for long-term storage in vacuum sealed aluminum foil packets. The seed vault, whose reinforced concrete walls are almost two meters thick, is set on a "floating" foundation that can absorb even severe earthquake tremors. It is kept at a cool thirty-five degrees Fahrenheit with a humidity of 17 percent. Electrical power is backed up by a primary generator, which is itself backed up in case of an emergency. Chang said that seed thus maintained can be kept a century.

Half a kilo of seed—500 grams—goes to short-term storage. The seed is divided into samples of 20 grams each, which are kept in small, brown-paper envelopes. This is IRRI's working collection, organized into "stacks" as in a library. Each variety has a number kept in sequential order in a basket on a shelf. When requests for seed samples come in, they can be blocked numerically. When the initial stock is down to 100 grams, an entry is made in the computer to restock. The short-term storage facility has a temperature of forty-five degrees Fahrenheit with 50 percent humidity. There, seed samples can be kept for up to seven years.

The short-term facility keeps enough seed on hand to meet the demand from breeders. In 1993, the germ-plasm bank sent out almost eleven thousand accessions to scientists in thirty countries as well as almost fifteen thousand to IRRI scientists.[90] To improve a rice plant, breeders have to select for the desired trait or cross the trait in from another plant. Breeders do not make genes, they have to find them. The trait must already exist somewhere in the gene pool for rice. That being the case, the more IRRI knows about the varieties in its collection, the more useful the collection is. Consider a search for rice plants likely to have drought tolerance. Breeders start with a computer search, looking for rices that already grow in a drought-prone environment. Hence, to be useful, the database has to have specifics about where a variety comes from. And files have to be updated continuously. When a variety is screened for a new trait, the information is added to its passport.

Collection

Breeding began at IRRI with rice germ plasm donated by the United States, Japan, Taiwan, and the FAO. By the end of 1962 the Institute had received almost seven thousand accessions from more than seventy countries. Thirty years later it had close to eighty thousand, mostly *Oryza sativa*, varieties, donated by more than a hundred countries. The collection is held in trust for all nations; IRRI does not claim property rights on germ plasm, nor does it patent new varieties. No country ever paid royalties for IR8. Through IRRI, rice germ plasm is distributed without restriction. Between 1986 and 1991, it sent out more than 270,000 rice packets to researchers around the world.[91]

The rapid spread of high-yielding semidwarf rices eroded the crop's genetic diversity. In irrigated environments, farmers abandoned traditional rices so quickly that many varieties were soon in danger of dying out. Changes in a production practice can have similar consequences. For example, escaping monsoon floods by switching to direct seeding puts older, deepwater rices at risk. So conserving rice germ plasm is a race against time.[92]

Dr. Duncan Vaughan, a gangly, sandy-haired Englishmen, organized IRRI missions for germ-plasm collection.[93] Working with national programs, IRRI scientists such as Vaughan have collected thousands of local rices all over Asia. Teams gather material for both IRRI and national collections. Seed samples usually come from farmers or are purchased at local markets known for their diverse selection. And there are rices for ritualistic purposes best obtained at the temples. In 1990, germ-plasm expeditions were underway in Indochina, where warfare had impeded collection for years. Germ-plasm collection also includes wild rices. Since 1989, more than one thousand samples have been collected on expeditions throughout South and Southeast Asia.[94] When I met Vaughan, he was preparing a mission to Papua New Guinea. For wild rice species, it is a place of extraordinary botanic richness where the flora of Asia and Australia meet. "Wild rices sprout up everywhere," he said, "in full sun, even in shade."

When obtaining a rice sample from a farmer, precisely what kind of data do scientists record? According to Vaughan, the stumbling block is figuring out what information is worthwhile. The knowledge farmers have about an old rice landrace may be the vital clue breeders need for the future.

There are many questions that local farmers can answer best.[95] Is a variety adapted to shifting cultivation? Is it grown on terraces? Can it be directly seeded? Can it be transplanted or double-cropped? What are a variety's typical yields in both good years and under adversity? How does it cook and what are its eating qualities? What do farmers consider a variety's main advantages and disadvantages? Unfortunately, thousands of rice varieties were collected before anyone paid attention to farmers' knowledge.

In Cambodia, damage to the country's rice inheritance is already extensive. In the terrible warfare of the 1970s, starving people ate the last seed stocks they had. Fortunately, IRRI scientists braved the civil war to collect seeds from farmers. In the 1980s, as the country recovered, IRRI restored more than five hundred traditional Khmer rices to Cambodia. They all came from IRRI's germ-plasm bank. Although all IRRI's accessions germinated, Cambodia had still lost hundreds of irreplaceable rice varieties.[96]

Without the full genetic diversity of *Oryza sativa* to draw on, adapting the rice plant to climatic change will be difficult, perhaps impossible. What impact will a depleted ozone layer and more ultraviolet radiation have on rice yields? How much does the methane gas released by flooded rice fields contribute to global warming? Both problems are under study at IRRI. To develop a rice plant that tolerates more radiation, or one whose roots release less methane, which genetic traits will rice breeders need? No one knows for sure. Without the right genes, biotechnology could be too little too late.[97]

PRODUCTION TECHNOLOGY

The core of IRRI's rice strategy is to develop high-yielding varieties with wide adaptability. A better plant, however, is just the start. To increase rice production, the difference between the yields researchers get and what farmers get has to narrow. That gap has much to do with how farmers protect their crops against weeds, insects, and diseases. New farming practices are often as important to high yields as a variety's genetic potential. How farmers do things, the agronomy involved, has an impact all its own.

"On experimental plots," notes Dr. Virgilio Carangal, "researchers get 7 to 8 tons. Most farmers are happy with 3 or 4. Closing the gap could provide much of the rice Asia will need in the next century." Carangal is from the Philippines. His task at IRRI is to analyze Asia's diverse rice-farming systems. Although irrigation creates the best conditions for high yields, millions of rice farmers are still dependent on the monsoons. About a third of this rain-fed area has such a dependable rainfall pattern that conditions are almost as favorable for rice as irrigation is. Still, this leaves two-thirds of the rain-fed area, more than 22 million hectares, at risk from periodic droughts and flooding.[98] Under such unfavorable conditions, changing how the rice crop is established, namely, from transplanting to direct seeding, holds great promise.

Direct Seeding

David Mackill, an American agronomist specializing in Asia's rain-fed lowlands, explained the change from transplanting to direct seeding.[99] Farmers transplanted rice for thousands of years and had good reasons, flexibility among them, for doing so. They can keep their seedbeds intact while they wait for the rains to fill up the paddies and, once filled, a flooded field kills most weeds. But there are drawbacks. When rice is transplanted, it has to be uprooted. It takes about three weeks for the plant to get reestablished, and even then, it never generates a root system that penetrates very deeply. As a result, transplanted rice is very susceptible to drought. When the rains stop and a field dries out, the losses are much greater than for directly seeded rice.

To plant rice directly means to broadcast the seed in a prepared field—as opposed to transplanting young seedlings into a flooded paddy. Direct seeding may be either "wet" or "dry."[100] With wet direct seeding, pregerminated rice seed is broadcast into wet soil. With dry direct seeding, dry rice seed is broadcast into soil with just enough moisture for germination. In either case, farmers save on water and on labor costs. Most important, they save time, because they can plant at the start of the rainy season, four to six weeks before they can transplant. Thus, farmers get the crop in before the heaviest monsoon rains come. With direct seeding, the root system is deeper and stronger, which helps the plant get through dry spells. When the heavy rains set in, the rice has deep enough roots, and is tall enough, to withstand flooding. With transplanting, if severe rains come before the rice can reestablish itself, it gets washed out.

Traditional rain-fed varieties are photoperiod sensitive; they flower and mature only as daylight hours decrease. Between the equator and China's Yangtze Valley, situated at roughly thirty degrees north latitude, the time to flowering increases greatly. In Bangladesh, the difference in daylight between the winter and summer solstices is more than two hours—and with a sensitive rice cultivar, fifteen minutes can have an impact. So even when farmers get a head start with the dry-seeded method, the time to harvest is still the same. And with photoperiod-sensitive varieties, that is a long time: between 190 and 240 days. The alternative is to substitute varieties insensitive to day length, a common attribute of semidwarf indica rices, which reach maturity in 90 to 100 days.

In Bangladesh and Southeast Asia, where rain-fed and deepwater rices are prevalent, direct seeding of semidwarf rices is preferable. Because traditional varieties take so long to mature, they are subject to damaging rains and high winds at the height of the monsoon season. However, with direct seeding, noted Mackill, "farmers can plant as soon as the first rains of the season come in late April—they do not have to wait for paddies to be puddled and saturated first. And they can harvest before the worst weather strikes in August." In many cases, farmers can plant a second crop after the monsoon subsides in September, thus benefiting from two high-yielding crops.

Localities sometimes prefer their traditional rices, or they want to plant at least some of their favorites. In such cases, direct seeding still has the advantages noted: it uses less water, saves on labor, and helps plants tolerate both dry spells and flooding. But direct seeding has its disadvantages. When dry seed is broadcast into a field with little moisture, weeds come up along with the rice—much more so than with seedlings transplanted into a flooded paddy. A dry-seeded rice field just getting started looks so much like common grass it is hard to tell the rice from the grassy weeds. Using a herbicide in conjunction with dry-seeded rice is the recommended solution. In the United States, where irrigated rice fields are directly seeded, herbicides are used on more than 90 percent of the acreage in production.[101]

In rain-fed zones, farmers claim that sowing at denser rates in wet fields helps suppress weeds. After germination, a few days without rain favors the more durable rice over many weed species. Nonetheless, with direct seeding, the weed problem is still a liability. In irrigated fields, farmers control weeds by keeping the water level in their paddies deep enough. In conjunction with wet-seeded rice, farmers with irrigated fields can eliminate transplanting altogether. To directly seed an irrigated paddy, the seed is first soaked for a day and then covered with a moist cloth for forty-eight hours. The pregerminated seed is then broadcast into the paddy, which is already puddled. As the rice sprouts, more water is channeled into the field. With this method weed control is comparable to transplanting and much less problematic than with dry-seeded rice.

Weeds

To narrow the yield gap, controlling weeds is high on the priority list. With direct seeding, weed problems usually get worse. Even with transplanting and irrigation, losses to intrusive weeds can reduce the total harvest considerably. Dr. Keith Moody, a stocky Australian in his late forties, leads the battle against weeds.[102] "Every year, every rice field gets weed problems," he said. "Losses from weeds are much greater than from insects, but compared to the lethal impact insects can have, weeds seem almost benign."

How do weeds end up in an irrigated paddy? When farmers establish their seedbeds, they broadcast the rice seed over moist ground in a small plot. Grassy weed seed from surrounding fields blows into the bed and germinates along with the rice. Later, when transplanting from the seedbed to flooded paddies, the weeds get transferred too, as they look deceptively like rice plants. By the time farmers can tell the rice from the weeds, it is often too late. If they pull up the weeds, they disturb or uproot nearby rice plants. If they cut the weeds back, they regrow like any grass. An herbicide does not do much good either, once the weedy grasses are established.

With transplanted rice, weeds are a nuisance. In a dry-seeded rice paddy, weeds can overtake a field before farmers have a chance to flood it. So far, the only solution is to work a selective herbicide into the soil before broadcasting the seed. Even then, the ground has to be moist and subsequent rainfall sufficient so that the herbicide gets absorbed by the soil. Once the rice germinates and its roots are established, fields can be flooded.

Direct-seeded rice grown under rain-fed conditions is the most weed prone. With a rain-fed paddy, keeping enough water in a field to drown the weeds is difficult. In the Philippines, it is not uncommon to have two or three weeks without rain in the middle of the monsoon season. Rain-fed fields can dry out, creating an ideal situation for hardy weeds.

"We know the best time and the best way to apply herbicides," Moody notes, "but they cannot be relied on exclusively. Even though herbicides break down rapidly, do not build up in the food chain, and are less toxic than insecticides, they still carry environmental risks." The runoff from herbicides contaminates groundwater, can affect drinking water, and can be toxic to fish.

Fortunately, there are ways to reduce herbicide use. Rotations from rice to other crops can kill off the most persistent weeds. After all, a grassy weed that mimics rice can escape both the trained eye and a selective herbicide. But when the next crop is mung beans, a popular rotation in the Philippines, looking like rice is a disadvantage. In fact, the reason many Asian farmers directly seed their rice plots is to gain time for a second, alternative crop. And after directly seeding a paddy, preparing the soil for a rotation is much easier. Although farmers still dike and flood their paddies, they do not have to puddle the soil into the kind of mucky paste needed for transplanting. As a result, fields are not so sticky, which makes them much easier to work.

For centuries, farmers selected rice plants for transplanting. According to Dr. Pablo Escuro, what is needed now is a plant adapted specifically to direct seeding.[103] A soft-spoken man with steely gray hair, Escuro headed up the Philippine National Rice Research Institute, or PhilRice. Trained at the University of the Philippines at Los Baños, he brought many years of experience to his post. In 1975, he joined an international team that helped Burma set up its own rice improvement program. By the time he returned to the Philippines in 1985, yields in Burma had increased by more than a ton per hectare.[104]

Escuro had worked for three years on an "ideotype" variety specifically for direct seeding. "A direct-seeded rice plant cannot be too tall," he said, "or it will be blown over. It cannot be too short, or the weeds will overtake it. The seed has to germinate rapidly and grow quickly so it can compete with weeds. We need a plant whose blades droop somewhat to inhibit weeds, but turn erect at flowering time for better light reception." It is the weed problem that determines the practicality of direct seeding.

Engineering

Seed broadcast by hand into a field makes weeding difficult. Inexpensive machinery like simple seeders that plant in straight rows made weeding easier and contributed much to Asia's rice revolution. According to Graeme Quick, a lanky, rambunctious Australian who headed up engineering at IRRI, small manufacturers all over Asia copied equipment originally designed at IRRI.[105] "We send out half a ton of simplified blueprints on request each year," he said. "In fact, our small tillers, seed planters, harvesters, and threshers were built so local craftsmen could pirate them even without a blueprint." In the Philippines, backyard welders and machinists turn out reasonable facsimiles of IRRI equipment. For preparing the soil, the power tiller is the most popular piece of equipment. In irrigated fields, it is used with pontoons on the back, so the tiller can float on the surface of a mucky paddy. It is pushed by hand from behind. There are attachments too, some designed to incorporate green manure ground cover into the soil.

The contribution small-scale mechanization makes is obvious in threshing, a time-consuming task usually done by women. To remove the rough rice from the panicles, the grain has to be pounded three separate times, the chaff discarded after each threshing. "Once the rice is cut," Quick said, "it is threshed in the fields and bagged right away. So farmers are afraid to cut too much at one time. Yet if they stretch the job out, the grain looses moisture, the panicles shatter, and the rice falls to the ground. A late harvest means low yields and poor grain quality. With hand threshing, when the monsoon rains continue into harvest time, it creates a terrible crisis." Mechanized threshing, by contrast, reduces waste, saves on time and labor, and gives farmers greater flexibility: they can harvest as much grain as they want when they want. The threshers Quick showed me were simply constructed and portable. To rotate the cylinders, some models use pedals and foot power; others have a small, five-horsepower engine.[106]

Postharvest Technology

A higher yield is one thing, processing the harvest is quite another. In the 1970s, postharvest losses in South and Southeast Asia were estimated at between 13 and 34 percent of the crop: during harvesting and threshing, 5 to 15 percent; during cleaning and drying, 2 to 3 percent; in storage, 2 to 6 percent; during milling, 3 to 7 percent; and during handling and transportation, 1 to 3 percent.[107] So there are good reasons for countries to target postharvest technology; it is basic to the success of a country's production program.

There are many steps involved in processing.[108] Once threshed, farmers have a stock of rough rice that has to be cleaned and dried. Winnowing was the traditional way to rid the rice of straw, stones, and inferior, poorly filled grains. Today, farmers often share small, mechanized cleaners that use vibrating screens and air blowers to remove the debris.

When fully mature, rice grains have a moisture content of at least 20 percent. To prevent deterioration and sprouting, the moisture must be quickly reduced to not more than 13 percent. Drying is done in the sun on mats, on concrete floors, and even on paved roadways. It can also be done in large drying bins heated by burning rice hulls, wood, or fossil fuels.

Once cleaned and dried, farmers sell their rough rice to local cooperatives, commercial buyers, or to government agencies. The next step is storage, in bulk or in bags. Facilities should be rainproof and airy with the floors sealed against moisture and rodents. Building materials include reinforced concrete, brick, wood, or sheet metal. Storage losses can be high, due to insects or poor moisture control. In the tropics, humidity is often 90 percent with temperatures in excess of eighty degrees Fahrenheit, so the moisture content of stored rice can easily increase beyond safe levels. The rice deteriorates, attacked by assorted bacteria and fungi. To prevent such losses, a good storage bin needs to circulate drier air, keeping the humidity at less than 70 percent.

The last step is milling. For a brown rice, the mechanized steel hullers remove only the outer husk. For a polished white rice, the bran layer is also removed. The final product is 80 percent starch and about 7 percent protein. Brown rice has a bit more protein and more B vitamins than its polished version. But it is less digestible and, consequently, less nutritious. Moreover, in hot, humid climates, the bran layer gets rancid. In Asia, almost everyone wants their rice polished.[109]

Recycled rice byproducts have many uses. Milling leaves behind tons of unused hulls and bran. The engineering division of IRRI wants to convert hulls into a high-quality fuel source that substitutes for firewood. It tried carbonizing them into briquettes for cooking, but the burning quality is low. Graeme Quick considered direct combustion more promising. That certainly is true in Louisiana. Agrilectric's 10-megawatt facility at Lake Charles burns the hulls to produce stream for generating electricity. Down-sized 2-megawatt versions of the power plant are already being sold in Asia. As to bran, it is rich in protein, vitamins, and trace minerals. Farmers mix it into rations for animal feed. A high-quality cooking oil can be extracted from bran; it can also be processed into flour and even used in soaps.[110]

Fertilizer

Modern rice varieties respond well to fertilizer. For example, IR8 was compact and high-tillering; when fertilized, it put out more panicles, which in turn set more grain. Nitrogen is the main component in chemical fertilizers. A rice plant, whether traditional or modern, needs 1 kilo of nitrogen to produce between 15 and 20 kilos of grain. In the tropics, the microbes associated with a hectare of rice naturally fix about 60 kilos of nitrogen, enough to set between 1 and 1.5 metric tons of grain. More production requires more nitrogen. The easiest and cheapest way so far has been from the fertilizer sack. According to Dr. Ronald Buresh, at IRRI on assignment from the International Fertilizer Development Center in Muscle Shoals, Alabama, "Almost 60 percent of the nitrogen fertilizer used in Asian agriculture goes to the rice crop."[111] Nitrogen for rice comes from the organic matter in the soil; it is also fixed by other plants and microganisms. Only at very high levels does chemical fertilizer account for more than half of the nitrogen used by rice. "The best strategy," said Buresh, "is to combine the use of each source in the field—organic matter from crop residues, biological fixation, and chemical fertilizer."

Rice needs less nitrogen than either wheat or maize, but how it is applied is inefficient. Moreover, most chemical nitrogen is based on petroleum, a nonrenewable resource with an uncertain price.[112] For countries without much oil of their own, which means most of Asia, reliance on petrochemicals is risky. For Asia's irrigated rice, granular urea is the main source of chemical nitrogen. Typically, the urea is broadcast over a puddled field, where much of it dissolves in the water and converts to ammonia. The nitrogen loss, according to Buresh, is at least a third and as much as a half. By contrast, inserting the granules deep into the muddy soil greatly reduces such losses. This is done at transplanting time, and only once, as it is very labor intensive. In Indonesia, deep placement is promoted as part of the country's rice policy.

A second nitrogen application comes just before the plant's seed panicles start to appear. By then, the rice has spread out and can take up the nitrogen quickly, which reduces losses to ammonia. Farmers broadcast the urea over their paddies. The timing has to be right, however, or the impact on yields is slight.

In Asia, most farmers harvest both the rice and the straw; there is little residue left to build up the soil. Adding chemical nitrogen can make up the deficiency temporarily, but eventually, insufficient organic matter and too much nitrogen weaken the soil's structure and yields fall. Consequently, efficient use is only one aspect of a larger problem. How can high yields be sustained if chemical nitrogen contributes less and less to productivity? And what if farmers cannot afford or do not have access to petrochemicals? Fortunately, there are sustainable ways to build up the soil's organic nitrogen.

"To make use of nitrogen, it has to be converted into a form the rice plant can take up," explains Dr. Jagdish Ladha, a young, articulate Indian scientist specializing in soil microbiology. That change is carried out by microorganisms, some of which need water to function.[113] Because rice is a semiaquatic plant, nitrogen can be produced in association with water-loving, nitrogen-fixing organisms. Azolla, a tiny, floating aquatic fern, fixes nitrogen symbiotically with blue-green algae. Farmers in South China have planted azolla in their rice paddies for centuries.[114] Ladha was enthusiastic about azolla. When soils have enough phosphorous, he said, and when temperatures are moderate—no cooler than sixty degrees Fahrenheit at night or warmer than eighty-five degrees during the day—azolla fixes more than 50 kilos of nitrogen per hectare. Its fronds spread out over the surface of the water, looking like small, wrinkled lily pads. Azolla thrives in saturated paddy soils, reproducing prodigiously by spores.

Farmers can establish azolla before they transplant their rices; all the azolla needs is soil sufficiently saturated so its roots can penetrate. For the rice to get access to the fixed nitrogen, the azolla has to be incorporated into the soil between the rice rows every fifteen days. The azolla is usually established first, then, during the first thirty days of the rice cycle, it is incorporated twice. The fixed nitrogen is released gradually—30 percent within five days, 70 percent within three weeks—so less is wasted.

In China, azolla is used as both a fertilizer and an animal feed. Once harvested, it is dried, ground up, and mixed into animal rations: up to a 30 percent azolla content in poultry feed, up to 50 percent for swine. Well-managed, azolla doubles in weight every two or three days. So farmers can harvest some daily without disturbing the balance in the azolla pool. In fact, both uses, as animal feed or as fertilizer, are compatible.

To build up organic nitrogen in their rice fields, farmers need more options than just azolla.[115] Early maturing semidwarfs and direct seeding give farmers more flexibility in planting and harvesting. Even when they double-crop rice, farmers have time to add a nitrogen-fixing "grain" legume (beans) to their rotations. In many cases, the best alternative is to plant the legume at the onset of the dry season, when the rains taper off and less water is available for irrigation. In the Philippines, the mung bean is particularly popular. Elsewhere in Asia, farmers prefer cowpeas, pigeon peas, or soybeans. All hold up well under drought stress, and after the harvest, farmers can incorporate the residues into the soil, improving its nitrogen content.

Legumes have direct access to nitrogen because of a symbiotic relationship with certain bacteria.[116] The carbohydrates the plants manufacture feed bacteria in the root nodules, which in turn fix nitrogen. Most grain legumes are edible, but there are many leguminous plants, such as *Sesbania rostrata*, that are not. A legume such as the soybean helps maintain and recycle nitrogen to the soil but adds less than the nitrogen-rich sesbania alternative. A remarkable plant identified in Senegal, sesbania has nitrogen-fixing nodules on its stem rather than on its roots. And it can grow in the standing water of a rice field. The taller the plant gets, the more nitrogen it manufactures. That means between 90 and 136 kilos of nitrogen per hectare—enough to justify its reputation as a "green manure," worth planting for its fertilizer value alone.

Why would farmers plant green manure instead of a food crop? According to Ladha, there are many reasons. Farmers can incorporate the sesbania into their rice paddies after just forty-five to fifty days of growth. It takes about ninety days to harvest many of the popular bean alternatives. Sesbania grows rapidly to a height of 1 to 2 meters, so it is well adapted to excessive rainfall and flooding. And if the problem is lack of rain, all it needs is a little moisture after germination. Of course, it has to be incorporated into the soil, which is extra work. To break down the nitrogen, sesbania has to be plowed under when fields are wet and muddy. For farmers without a hydrotiller, there is a prior step: they have to cut back the sesbania stalks before they plow. In fact, without the hydrotiller, Indian farmers would not have adopted sesbania so fast. "The hydrotiller," Ladha said, "reduced the time needed to puddle fields, making it possible to add a sesbania crop to the cycle without adding to labor costs."

"Increasing the soil's organic nitrogen," Ladha said, "makes sense to researchers, but it does not always seem that way to farmers." Planting sesbania is additional, labor-intensive work. So Ladha surveyed farmers in Gujurat, India, to find out why they planted it. The main reason cited was depleted soils, which led to yields that kept falling despite adding chemical fertilizer. With sesbania, farmers found they could cut back on chemical nitrogen by half.

What does *Sesbania rostrata* look like? At IRRI's research site, the sesbania was only two months old, but it was 3 meters high. The stalks were the size of giant sunflowers, although much woodier and stronger. In clumps, the sesbania looked more like a gangly barrier of bushes than anything remotely akin to a bean plant. In India, farmers sometimes cut back the woody stalk as firewood for cooking. Tender shoots and leaves can be ground up and fed to animals—of course, this reduces the biomass and, hence, the nitrogen added to the soil.

To make sesbania worthwhile for farmers, timing is crucial. Where double-cropping is the norm, only about thirty days remain for a third crop. That is insufficient for sesbania. Farmers, however, can broadcast sesbania seed into their fields a couple weeks before the rice harvest, thus giving it a head start. And even though it gets trampled somewhat during the harvest, it recovers and resumes its rapid growth. Incorporation takes place before planting the next crop.

For many farmers, adding another crop rotation is not practical. *Aeschynomene afraspera*, a bushy, leguminous plant from Madagascar, is shorter than sesbania and has softer stems, making it easier to incorporate. I visited trial sites at IRRI with Ladha. The advantage afraspera has over sesbania is that farmers can plant it along with their rice crop. They end up transplanting the same number of rice seedlings; what changes is the geometry of the paddy. Instead of neat, equidistant, transplanted rows, farmers plant compact double rows with enough space left for the afraspera inbetween. Both grow together for about five weeks, then the tender legume is incorporated by simply stamping it into the soggy ground. Trial results show healthier, more vigorous rice plants that set grain exceptionally well. And a healthy plant better withstands the diseases and insect pests that attack it.

Plant Hoppers

After paging through IRRI's handbooks on pests and diseases, manuals that are published in more than twenty languages, it seemed a rice plant had to beat the odds to produce anything. Once past the leaf cutters, stem borers, and plant hoppers, it still has to survive blast fungus, tungro virus, and bacterial wilts. "Rice has been planted contiguously and continually on millions of hectares for thousands of years," noted Dr. Kong Leun Hoeng, a fast-talking entomologist from Malaysia.[117] "This helps explain why so many monophagous pests depend on it to survive. With most crops, the insects that attach it are polyphagous—they can eat many different plant types, not just the crop in question."

For gardeners in North America, the omnipresent aphid is a classic example of a "polyphagous" insect. It can cause trouble for almost any leafy vegetable, from tomatoes and peppers to lettuce and pole beans. And the flower garden is not immune either, especially the roses. As a result, a tomato plant with resistance does not put the aphid under selective pressure—it just moves on to other plants. That is not the case with a monophagous rice pest like the plant hopper. Its reproduction and eating habits are geared exclusively to rice. Plant hoppers suck nutrition from the plant's stem and then inject their eggs. In Asia, plant hoppers have done tremendous damage. Consequently, IRRI breeders are constantly looking for sources of resistance. Eventually, they found a plant that carried a toxin; when plant hoppers fed on the plant, most of them died. The trait was crossed back to improved varieties, but the resistance did not hold up for long. When breeders find resistance to a soil stress such as salinity, the soil does not fight back. Insect resistance is very different, particularly with a pest as specialized as the plant hopper. A resistant variety puts the insect under extraordinary selective pressure. It fights back with population genetics.

Like many insects, the plant hopper has a survival strategy based on high rates of reproduction. "Such insects," Hoeng noted, "are usually small, they can migrate long distances, have a short life cycle, and can lay thousands of eggs." A plant hopper reproduces in just forty days—eighteen days from larval stage to hatching. Once hatched, it lays from forty to eighty eggs per day for twenty-two days. The population is composed of different biotypes. Under selective pressure, the distribution shifts. A resistant, toxic rice plant can kill most of the plant hoppers, maybe even 98 or 99 percent. But the biotype that survives can feed on the rice despite the toxin. "Eventually," Hoeng emphasized, the plant hoppers will make a comeback."

Indonesia is a case in point. Early semidwarfs, such as IR8 had no plant-hopper resistance. A disastrous outbreak in 1976 destroyed hundreds of thousands of hectares, despite the heavy use of insecticides. Although IRRI had a resistant semidwarf on hand, IR26, it soon fell prey to new plant-hopper biotypes. More durable resistance was provided by IR36, but eventually, insect mutations led to new outbreaks.

In its fight against plant hoppers, Indonesia combined resistant varieties with heavy insecticide use, in this case, a self-defeating strategy. Hoeng explained that when farmers apply insecticides, they often spray too early—when they see a lot of plant hoppers. This kills off the adult pests but not their eggs. When the eggs hatch, the problem gets worse, because insecticides have killed off plant hopper predators, particularly spiders. "The spider," Hoeng explained, "is bigger, moves around less, lays only ten to twenty eggs, and takes three months to reproduce. When its numbers are greatly reduced, as happens with widespread insecticide use, recovery takes much longer than for a wide-ranging, high-reproductivity pest like plant hoppers." The consequence is an upward spiral of insecticide use that reduces but does not prevent severe plant-hopper outbreaks.

Rice has adapted to insects, even to the notorious plant hopper. A plant can tolerate ten plant hoppers, but one hundred do serious damage and five hundred can suck the plant dry and destroy it.

The problem in Indonesia is not the plant hopper per se, but rather a rate of population growth unchecked by plant-hopper predators. In 1985, insecticides were unable to contain a new "brown" plant-hopper biotype. Indonesia faced a loss of a million tons of rice. With IRRI assistance, resistant varieties were identified. In the meantime, Indonesia changed its control strategy, cutting back on insecticides and opting for a more integrated approach to pest management. Farmers were trained in special field schools to gauge the severity of the plant-hopper problem. When the numbers did not exceed defined thresholds, they relied on predators to keep the plant-hopper population under control. In pilot projects, insecticide use, which was heavily subsidized, fell by almost 60 percent. Subsequently, farmer field schools and integrated pest management (IPM) became state policy for rice.

The government banned fifty-seven types of pesticides. Plant hoppers can still be sprayed as a last resort, but the pesticide in use does not kill its predators.[118]

"The only way we can get rid of plant hoppers," Hoeng said, "is to get rid of rice. Pesticides can win some battles, but they will never win the war." That being the case, coexistence is the best strategy. Instead of a total war against plant hoppers, "we need a rice plant with greater tolerance to attack," Hoeng said, "a plant whose enzymes reduced egg viability, thus keeping growth rates manageable".[119]

Pesticide use in Asia accompanied the spread of modern rice varieties. For example, IR8 had little built-in pest resistance. To protect their crops, farmers were told to spray on a regular basis. And they still do, even though subsequent varieties had much more pest resistance. Farmers in India, for instance, apply about a third of a kilo of pesticides (active ingredients) per hectare of rice; the average in the Philippines is about half a kilo. This is not much compared to Japan and South Korea, where rice farmers used a staggering 14.3 and 10.7 kilos of pesticides (active ingredients) per hectare, respectively. In the United States, farmers use pesticides on about 12 percent of the country's total rice acreage.[120]

Even in small amounts, pesticides create serious problems, especially in Third World countries. In the Philippines, farmers spray two or three times a season. They do so on a calendar basis unrelated to either actual pest densities or potential yield loss. They spray pest-resistant modern varieties just as much as nonresistant ones. And as noted, indiscriminate pesticide use disrupts the pest-predator balance, leading to subsequent insect resurgence and yield losses. Philippine farmers commonly use pesticides the World Health Organization (WHO) classifies as extremely or moderately hazardous. Moreover, the way farmers mix, apply, store, and dispose of pesticides is often unsafe. A study based on a random sample of rice-farming households in the Philippines showed that farmers who applied three insecticide doses per crop had a one in two chance of developing chronic eye problems; they also ran higher risks of skin ailments and respiratory abnormalities.[121]

Thus there are many reasons IRRI now promotes an integrated approach to crop protection. The strategy involves complementary practices. Farmers are encouraged to rotate crops, to time the planting cycle so it minimizes pest attacks, and to rely on biological control with pest predators. Pesticides are viewed as a last resort, applied only when insect densities threaten serious yield losses and at the minimal concentrations necessary.

Plant Protection

"Building resistance into the genetic armor of the plant has its benefits," noted Dr. Ramesh Saxena, an entomologist from southern India. An earnest, practical man in his fifties, Saxena felt that IRRI's research was often too theoretical.[122] "A variety that is resistant to a prevalent pest or disease," he said, "benefits all farmers equally. Integrated pest management, by contrast, can be overly sophisticated and impractical. Farmers are supposed to monitor pest levels in their fields. But they get tired of taking samples; they already have too much to do."

Releasing resistant varieties encourages farmers to switch over and replenish seed stocks. According to Saxena, unless a better variety comes along, farmers recycle the seed from crop to crop. The quality of such seed eventually deteriorates. Weed seed gets mixed in, or there is cross pollination from different varieties nearby, which reduces seed purity. Fungus diseases can be a problem too, especially if the seed is not treated. For the best yields, starting out with high quality, disease-free seed is crucial.

Insect losses are not just confined to fields. Once harvested, rice has to be stored, both for milling and, in many cases, for seed stock. In hot, humid climates, keeping storage bins free of insects is particularly difficult. "If we paid more attention to what farmers do," Saxena said, "we might find simple solutions to problems." In southern India, for example, high humidity creates conditions conducive to insect damage during storage. Losses can be from a third to half of the rice. Fortunately, Indian farmers have devised ways to protect their rice stocks. They take leaves from the neem tree, a locally abundant native species, and line the storage area with them. The leaf has a natural chemical that insects avoid. Turmeric, a common root spice and dye, is another example. "Mixed in with the rice," said Saxena, "turmeric both protects the grain in storage and improves cooking quality." Is it not better, he asked, to use neem leaves and turmeric rather than chlorine based chemicals?

Saxena helped bring neem trees to the Philippines, where IRRI has distributed more than seventy-five thousand seedlings. Its uses and application are many. When ground up and mixed with urea nitrogen granules, rice yields increase by half a ton per hectare. In India, neem cakes are sold as fertilizer. Neem also makes an effective, biologically safe insecticide. Studies show that neem compounds repel insects, inhibiting feeding, reducing fecundity, or increasing larva mortality. In the United States, the Environmental Protection Agency (EPA) has approved neem-based products for use on both feed and food crops. Saxena said that neem is effective against the insect carriers of tungro virus, the most dangerous rice disease in the tropics. When mixed with cowpeas and mung beans during storage, it reduces insect infestations. Neem also has important medicinal properties. Neem oil, for example, suppresses several pathogenic bacteria and is toxic to fourteen common fungi. In India, neem is used as an antiseptic in soaps and toothpastes.[123] "In the tropics," Saxena said, "use of the neem leaf is almost cost free. Protection is not 100 percent, but it does not have to be. The technology is simple and inexpensive."

Tungro Virus

Modern rice varieties have had a tremendous impact on agricultural practices in Asia. Given the scale of the change, it is not surprising that the rice revolution has its negative side. According to Dr. Paul Teng, a young plant pathologist, the spread of tungro virus is an example.[124] To take advantage of modern rices, Asian countries invested heavily in irrigation. Because modern varieties mature so rapidly, farmers can plant a second rice crop. Sometimes they even manage five rice crops in two years. Without a nonrice rotation staggered in, it is easier for rice insects to survive from crop to crop. According to Teng, a U.S. citizen born in Malaysia of Chinese ancestry, "Several different insects can transmit tungro virus. Planting rice after rice builds up the carrier population, resulting in severe outbreaks of the virus."

To reduce tungro virus, farmers should leave the land fallow or rotate crops. In Malaysia, a month-long fallow period between rice crops virtually eliminated the virus. By contrast, rotating to a different crop is much more work. Prior to the rice harvest, paddies are usually drained. Once exposed to the sun, the soggy soils become brittle and cracked, making them difficult to plow. So farmers resist new rotations, planting rice after rice as much as possible.

Sheath blight is a fungal disease, almost as serious in Asia as tungro virus. In China, Paul Teng estimated losses at 30 percent, which suggests it is a problem almost everywhere in the country. China irrigates almost all of its rice and relies almost exclusively on modern varieties. Although it developed dwarf varieties independently, almost all today have some IRRI ancestry.[125] "In the Chinese view," Teng said, "the connection with modern varieties is that the fungus has spread with the intensified use of nitrogen fertilizer." For irrigated rice, Asian farmers often apply between 150 and 200 kilos of nitrogen per hectare. Such heavily fertilized plants put out more foliage. Because modern cultivars are short, tiller prolifically, and are so densely planted, the concentration of leaves and stems creates a hotter, more humid environment with less air circulation--conditions ideal for fungus growth. When farmers apply 180 kilos or more of nitrogen but no fungicides, losses per hectare are about 1.5 metric tons of rice.[126]

For sheath blight, scientists have yet to find a genetic source of resistance. The remedies, as for tungro virus, are "cultural," that is, they depend on how farmers manage their rice crop. "An increase in the spacing between plants," said Teng, "or a more efficient use of fertilizer, could control the disease well enough." When farmers cut back on nitrogen to 100 kilos or less, the sheath blight infection rate is much reduced and losses from the fungus are negligible.[127]

Water

"If there is one thing we know about rice," Peter Jennings once told me, "it is that rice loves water." He ought to know. Jennings led the IRRI team that developed the first of the modern varieties that so changed rice production in Asia. Later, he moved on from IRRI to Latin America. When I met Jennings in 1983, he headed up the rice program at the International Center for Tropical Agriculture (CIAT) in Cali, Colombia. By the time I reached IRRI in 1990, Jennings had quit. But I remembered his lesson. Whenever I thought about rice, I thought about water. So did Dr. Sadiqul Bhuiyan, an agricultural engineer from Bangladesh. He works at IRRI promoting small-scale irrigation systems for local use.[128]

High-yielding modern rices need good water control. In much of Asia, given the monsoon deluge and the many river basins and deltas, it was not difficult, or particularly expensive, to convert to irrigation. Eventually, marginal areas were irrigated too, but expensive, state-run systems are many, with results that fall far short of expectations. Enough water at the right time is essential in rice production. In rain-fed areas, collecting surface runoff in ponds is an affordable alternative to expensive irrigation projects.[129] Small-scale systems are very productive, particularly when farmer-managed. According to Bhuiyan, almost 80 percent of the increased rice output Asia needs can come from such inexpensive, locally owned water supply systems.

Why do farmers want to collect water? In most of Asia, they face a wet season with floods followed by a dry season with water shortages and droughts. Without a backup water supply, it is risky to try for a second crop, even when they rotate to less water-intensive alternatives such as mung beans or maize. "With a small pond," said Bhuiyan, "farmers get a second crop during the dry season that supplements the rainy season crop. But they have to get the first crop out and the second crop in as fast as they can."

When the dry season starts, farmers calculate how much water they have relative to what different crops need. And they must figure in evaporation. They may have enough water for rice, but not for as much as they planted in the rainy season. On the land that remains, they can rotate to popular leguminous alternatives (mung beans, lentils, soybeans, peanuts), to maize, or to a green manure.

A pond takes up an average of about 7 percent of a farmer's land. A gravity-fed system is the least expensive, as water is collected and stored on upper terraces for distribution to lower ones. With flatter land, the water has to be pumped out of the pond. The main cost is constructing the pond and grading the area, usually by bulldozer. Bhuiyan estimated that total expenses for a pond of 1,000 square meters came to about four hundred dollars. Most farmers get back 60 percent of their investment the first year. Miscalculation is the main problem, planting more rice than can be irrigated. But farmers learn to work out estimates accurately, and in many cases, they used too much water. Rice does well enough in saturated soil moist enough to prevent drying or cracking. Although yields will not be as high, farmers can reduce water use in their paddies by 30 to 50 percent.

Because flooding fields helps to kill weeds, when farmers use less water, they end up with more weeds. Nonetheless, saving water in ponds can increase rice production in rain-fed areas subject to flooding and dry spells. A pond provides a reserve farmers draw on in periods of drought. Trying to breed for such extremes, by contrast, is difficult and time consuming.

"We have yet to breed a rice specifically for marginal rain-fed areas," Dr. Keith Ingram told me.[130] An agronomist from the United States, Ingram worked on breeding more drought-tolerant varieties. "Modern varieties were developed with irrigation in mind," Ingram said. "They are short with shallow root systems of less than 15 centimeters. So they are especially sensitive to drought." When transplanted to a rain-fed paddy, a three-week dry spell does a lot of damage. If the paddy dries up, even for a few days, the yield loss can be 50 percent. Upland rice, by contrast, has a much deeper root system—more than 60 centimeters in sandy soils. But upland rice does not like standing water; a flood will kill it. "What we need," Ingram said, "is a plant with the root system of upland rice, but which does well in saturated soils." Such a plant was a long way off. To get data on the depth and density of a particular plant's root system means digging up and around it for measurements when it reaches maturity; there are no short cuts. "For rainfed conditions," Ingram said, "we still lack a good ideotype, a sense for what the best root system would look like."

ON-FARM RESEARCH

Asia has to increase rice production now; it cannot wait for breakthroughs in biotechnology to save the day. New ways to seed fields, protect crops, improve the soil's fertility, and use water efficiently are solutions farmers can apply immediately. But do such schemes work as well for farmers as for researchers? No matter how good the research seems, its utility has to hold up outside the laboratory.

To make sure its research on agronomy is relevant, IRRI does "strategic" research on local farming practices. Of course, it cannot involve itself directly in thousands of extension projects all over Asia. Instead, it collaborates with national programs on carefully selected local projects. In 1993, IRRI had twenty-seven scientists posted outside the Philippines in twelve different countries.[131]

Dr. Virendra Pal Singh, a stocky, pragmatic Indian agronomist, works at IRRI with on-farm technology-transfer projects in eastern India and in the Philippines.[132] "To a researcher, a new practice seems clearly beneficial," said Singh, "but to a farmer, it is usually not so obvious." The on-farm projects identify what promotes and what inhibits the use of better varieties, different crop rotations, and new agronomic practices. To do this, they work closely with extension personnel. The objective is to design trials that involve farmers and that demonstrate the potential of a new technology. What they transfer is not just a solution but an approach that is farmer-oriented.

Incorporating a sesbania green manure into the soil is a case in point.[133] It is much easier to just add more chemical fertilizer. Many farmers resist sesbania because adding to the soil's organic content only pays off in the long run. Poor farmers focus more on immediate advantages. "Sesbania," Singh said, "has to provide benefits farmers value." Its woody branches, for example, can provide thatch and cooking fuel; the leaves make good fodder for animals, and when sesbania is planted around fields, it is a good windbreak. In paddies also used for raising fish, the strong stalks can make a kind of enclosure. In short, what makes sesbania attractive to farmers is often different from what researchers think is significant.

In dry soil, incorporating sesbania is particularly difficult, especially for farmers without a motorized tiller. It is often better to cut the sesbania down and make a compost pile. The following year, when the rainy season starts, the decomposed debris can then be spread over the field and easily incorporated. Farmers accepted this easily, as the practice is similar to collecting, composting, and incorporating animal manure.

That a new technology congruent with accepted practice spreads more rapidly seems obvious enough. That the research itself ought to begin with on-farm problems is more controversial. According to Dr. Sam Fujisaka, a Japanese American anthropologist, "Farmers involved in projects spend too much time on technology somebody else has decided is important."[134] Fujisaka's work at IRRI looked at farming from the farmer's point of view. "We need to focus more on what farmers are doing and why," he said. They used neem leaves, burned rice hulls, and raised azolla long before IRRI existed."

In fact, for many production problems, both IRRI and national programs could save a lot of time if they listened to farmers from the start rather than consulting them last. That is in part the rationale behind on-farm research: to learn from farmers how they establish and protect their crops. In Claveria, a Philippine township where Fujisaka did on-farm research, rice is the main crop. Farmers follow rice with a fallow, or try to fit in a maize crop, even though it is risky. What they lacked was the technology to make it possible. "If we can show them how to work in the maize successfully," said Fujisaka, "they will adopt the technology immediately; transferring it will not be a problem. Open a door for farmers and they will walk though it."

Santa Barbara

Do new varieties and cropping patterns make a difference? Do farmers walk though the door when it opens? That is the question that brought Tom Tengco and I to Santa Barbara in Pangasinan Province.[135] It is a rain-fed rice zone of diked fields. The average farmer has about 1.5 hectares of land, subdivided into parcels. Sharecropping or leaseholding are common. When farming on shares, 25 percent goes to the landowner, 75 percent to the producer. When the rainy season starts in May, farmers establish their seedbeds. Transplanting to paddies takes place between late June and early August, depending on when the monsoon rains set in. The harvest takes place between October and November. On-farm research at Santa Barbara involved the Philippine Department of Agriculture, the University of the

Philippines at Los Baños, and IRRI. The project documented local farming practices and introduced new production technology with the assistance of farmer collaborators who participated directly in the project's work.

Adding a mung bean rotation is an example of a local farming practice examined in on-farm research. After rice is harvested, farmers leave their fields fallow. But with only minimal soil preparation, they can get a mung bean crop in quickly before the rainy season ends. As Tengco said, mung beans have the kind of advantages farmers and their families appreciate: "They taste good, they are productive, and they are easy to plant." Yields are at least a ton per hectare, so there is plenty left over to sell in local markets. Once the beans are harvested, the plant can be cut back for animal fodder. The stubble that remains is incorporated, adding nitrogen to the soil.

Storing mung beans, however, is a problem. Insect damage is very high; at Santa Barbara, it was difficult to save enough beans to seed the next crop. So the project tested traditional control methods that the community's Women's Association suggested. Beans were kept in sealed plastic containers with various spices and leaves mixed in. The results were excellent. After three, six, and even ten months, losses were less than 2 percent for beans stored with garlic cloves or *madre de cacao* leaves.

For the Santa Barbara area, the project documented which tasks in rice production fell to the men and which to the women. The men establish the seedbed, women bundle the seedings, and men usually do the transplanting. Harvesting and threshing are joint activities. Making the sweet, glutinous rice cakes for which Santa Barbara is famed is a cottage industry the women manage. The delicacy requires an exceptionally sticky rice that can be harvested early and processed easily, criteria few modern indica varieties meet.

Working with local farmers, the project identified a modern variety that matured rapidly and was glutinous enough to pass the baking test. "Once introduced," Tengco said, "yields increased from 2.7 tons per hectare to 4.7 tons. And since it matures so rapidly, farmers have more time for rotations." The variety displaced had required at least four months to mature. The new variety takes just 110 days. So when the first rains come in May, farmers have time to plant a green manure such as sesbania or a rapidly growing forage legume. The payoff for farmers is fixed nitrogen for incorporation into the soil and a plentiful supply of animal fodder.

Feed for animals is important in its own right. For tilling fields, many farmers depend on the water buffalo, and most households keep a couple of cows, some pigs, and a few chickens. So adding a forage crop to rotations was an innovation that spread rapidly. Farmers sometimes combine the new technologies, intercropping mung beans with various forage options. Rice straw is an obvious forage source too. The problem is how to store it during the rainy season. The solution is to pile it up in huge stacks. When it rains, only the surface gets saturated. Farmers pull out the straw from the bottom of the stack, and it is always dry.

My orientation was not complete until I talked with a rice farmer. We set out in search of "Sarge" Garcia, who had returned to farming after retiring from the Philippine armed forces. We cut across a maze of earthen dikes and canals. Stacks of straw were piled under shade trees. Lucaena trees lined the main dikes and the road into town. Rapidly growing and leguminous, farmers cut back the tree's limbs for firewood and the leaves are used as animal feed. After trimming, lucaena recovers rapidly. Around Santa Barbara, we rarely saw a country road without it.

His face wrinkled and leathery, Sarge had spent the day transplanting. "The time to plant came early this year," he said. "I transplanted my first parcel three weeks ago on July third. The timing was very good, as the seedlings were just mature enough to plant. Last year, the rains stopped after I had started transplanting. The field dried up. I lost the part transplanted plus the fertilizer I applied. I did not get the rice in until August eighth. When transplanting gets delayed, the roots on the seedlings start to rot, ruining the seedbed."

This year, Sarge had planted six tenths of a hectare in rice. Last year it was less: four and one-half tenths. The parcels are worked by himself, his son, and his wife. Labor exchange between families is common. Like many small farmers in the vicinity, he hires a tractor driver to do the plowing and harrowing. Last year, Sarge harvested forty-two sacks of rice weighing 46 kilos each—which averaged out to more than 4 metric tons per hectare. At harvest time, buyers go from farm to farm purchasing rough rice. It is sold already threshed, dried, and bagged, which imposes an additional cost on farmers. The same people who control the tractor business also own the threshing equipment. "It's a monopoly," Sarge said, "but we can't hold out for better prices without a place for storage." To compete with the monopolies, local farmers had formed a cooperative. "We already have our own thresher," Sarge said. "A tractor is next on the list."

What had the Santa Barbara project accomplished? To Sarge, the most important thing was the increase in yields. Like his neighbors, Sarge stacked his rice straw and planted mung beans. After the bean pods were harvested, Sarge let his cow graze in the field. Just before planting rice, he incorporated what was left of the mung beans.

Sarge took us to his home for soft drinks. His house had cement-block walls and a high ceiling with a tin roof. There were terrazzo floors and for cross ventilation, jalousie windows. The family's food is prepared in a separate cooking area at the back of the house in a small oven that uses cow dung for fuel. Flowers and fruit trees filled the front yard, in the back was the barn, a pigsty, and an outhouse.

Guimba

To reach Guimba we headed south through Gamiling and then northeast to Nueva Ecija Province. At Guimba, farmers irrigate the rice crop. The terrain is too flat for a gravity-fed system. Instead, there are electric-powered pumping stations fueled by petroleum. When the system was constructed in the 1970s, energy was cheap, so farmers did not worry about how much water they used. By the early 1980s, however, the cost of electricity had risen so much that water use had to be restricted. The main problem now is how to diversify the cropping system and use less water.

Work at Guimba was a joint effort between the Philippine Ministry of Agriculture and IRRI. I met with the project coordinator, Hermenegildo Gines, a burly, tough-minded agronomist who had worked with the project since its inception.[136] According to Gines, IRRI had researched cropping systems it felt would help farmers. The Guimba project was a chance to evaluate the new technology. They began with a workshop that involved farmers, researchers, and local extension workers. The job was to diagnose the problems, outline the alternatives, and evaluate the results.

Guimba farmers plant the first rice crop in June, when the rains are steady, and harvest in October. In January they plant a second crop, which they harvest in May. The most serious problem farmers face is that less water is available for rice during the dry season. For every 100 hectares of rice planted in the wet season, there is only enough water to irrigate 35 during the winter dry season. And yields are much lower too: less than 3 metric tons per hectare compared to almost 5 tons during the rainy season. Even with ideal inputs, and even when they moved back the planting schedule for the second crop to November or December, the results were still disappointing. Why? "Because during the winter it is cooler, it is windier, and solar radiation is less," said Gines. "No matter what farmers do, yields from the second rice crop are only half of the first.

For farmers who want to plant rice twice, the solution is to escape winter weather altogether. To do this, farmers have to switch to directly seeded rice.

For two crops the best solution is to dry seed the first one in April for a July harvest. That leaves time to transplant a second crop from seedbeds in early August for a harvest in late November. "The great change in rice production here," Gines said, "is the flexibility farmers get by direct seeding. With transplanting, farmers have to wait for the monsoon rains to come. When they are late, which is not infrequent, they end up transplanting after the seedlings have passed their prime. With direct seeding, farmers do not puddle the soil. So the plant gets a deeper root system. The rice is already well established when the monsoons hit, so it withstands flooding better. Compare the directly seeded rice around here with the transplanted rice—it is three times the height and almost ready to flower." I did; he was right.

Guimba farmers who stay with transplanting do best by rotating to mung beans or maize for the second crop. Mung beans do not mind the winter weather. And compared to rice, mung beans use much less water. Farmers can substitute 2 hectares of mung beans for every hectare of dry-season rice they irrigate. With maize, however, water efficiency is not as great as with mung beans. For every hectare of rice irrigated, farmers can substitute only 1.5 hectares of maize. Nonetheless, because maize produces a high-quality animal feed, it fetches a very good price. In fact, it is more profitable than rice. As the maize rotation spreads, Guimba farmers are adding hog production to their range of activities.

Whether farmers double-crop with rice or rotate to a different crop, at the end of the cycle they still have time for a green manure. There are many options besides sesbania. The project relies on its farmer collaborators to identify the most suitable. For example, a slow-growing legume can be intercropped with maize for incorporation into the soil after the harvest. The important thing is to base such work on farmer participation and criticism.

TRAINING

The International Rice Research Institute worked on a genetic map for rice and wide crosses, it researched how best to apply fertilizer, how to reduce pesticide use, new cropping patterns, and weed control. How did the results get circulated to other countries?

In Asia, rice research is a shared responsibility. When IR8 was released in 1966, most Asia nations already had their own rice research programs. Today, half of IRRI's scientists are from developing countries. Dr. M. S. Swaminathan, a plant geneticist and rice breeder who had headed up the

Indian Council of Agricultural Research, was IRRI's director general from 1982 to 1988. Two research consortia have IRRI as a partner: for rain-fed rice in lowland environments and for work on upland rice. The institute coordinates networks for genetic evaluation (INGER), for research on farming systems, for integrated pest management, and for biotechnology. To benefit from seasonal differences, it does shuttle breeding with national programs in South Korea, Egypt, China, and Thailand.[137]

In 1992, IRRI sponsored conferences, workshops, and collaborative research meetings involving more than a thousand participants from forty countries.[138] It has designed and published primers on rice production, pest control, plant diseases, weeds, and organic-based rice farming. Such practical manuals are used by teachers, agronomy students, and local extension workers. And IRRI has an ingenious approach to translation and co-publication. As a result, its handbook, *A Farmer's Primer on Growing Rice*, is available in thirty-eight languages. It is a best seller, probably the most influential and widely used agronomy book ever published.[139]

Over the past thirty years, more than 7,000 agronomy students, research scientists, and production engineers have passed through IRRI to take courses, attend a workshop, or for hands-on experience with a new technology. In 1993, 256 scholars and postdoctoral fellows did collaborative research at IRRI. They came from twenty-nine countries in Asia, Africa, and the Americas. An additional 147 trainers participated in ten short courses held at IRRI headquarters.[140]

When IRRI's training program started, it concentrated on production technology for modern rices. National programs, particularly in Asia, now provide much training on their own. In 1993–94 thirteen courses were run in-country in Thailand, Vietnam, Bangladesh, Indonesia, and the Philippines. The institute's rice production research course was held at Thailand's Rice Research Center by a team of Thai and IRRI trainers. However, IRRI still provides specialized, in-service training. Its approach is focused and clear-headed. To diffuse a new technology, IRRI trains a core group in its use. In 1993, it held ten short courses on topics ranging from rice biotechnology and hybrid seed production to integrated pest management and weed control. Courses are practical; half the time is devoted to hands-on experience. Because it has such a good track record, IRRI now trains trainers from all over Asia.[141]

Because participants' backgrounds are diverse, courses are usually taught in English, the best common denominator. Nonetheless, for both teachers and students, English is often a second language, which makes communication difficult. The institute had to develop self-help materials so students could follow the courses, which meant booklets on rice production, on farming systems, and even on statistics. Its modules on rice technology are used in agricultural schools all over Asia. And for special courses, like pest management, it develops slides and tapes. In 1993, thirteen of its slide-tape modules on farming systems research were translated into Chinese.[142]

The importance of IRRI's training capacity is demonstrated in Cambodia. The country's internal strife had destroyed its research facilities, scientific equipment, and irrigation systems. Cambodia's agricultural specialists either fled or were killed. In 1965, the country had 2.3 million hectares in rice; in 1974, it had only 500,000, and in 1979, only 800,000. Cambodia had to rebuild its agriculture.[143]

In 1987, with support from the Australian government, IRRI began a crash program to boost Cambodia's production. The project works on new varieties, better production technology, and training. By 1991, it had trained more than one hundred advanced students in rice production. More important, it taught Cambodian scientists to organize their own production courses. Now, basic training is centered in Cambodia. The course is in the Khmer language—along with the manuals, slide modules, booklets, and primers.[144]

LESSONS

"To keep Asia's rice bowl filled," said Dr. Klaus Lampe, IRRI's director general, "good science is not enough." A German agronomist with a specialization in agricultural engineering, Lampe headed up IRRI from 1988 to 1995, a difficult period of budget cuts and staff reductions. "When scientists get too caught up in research papers," he said, "farmers and production get lost in the shuffle."[145] True enough. Complaints that IRRI had lost contact with farmers were frequent, often impassioned. The challenge is to recognize that some problems can be solved in the laboratory but others require collaboration with farmers. The choice between biotechnology and on-farm research is a false one. To feed a hungry planet requires both.

Mapping the rice genome is an obvious example of appropriate laboratory research, but much of IRRI's routine work is done without involving farmers, and rightly so. Examples are many. Before new rice lines are released, IRRI routinely screens them against pests, diseases, and different abiotic stresses. It coordinates INGER, which assembles rice nurseries for testing all over Asia, a task that requires IRRI to multiply and distribute seed for thousands of varieties each year. The institute's germ-plasm bank has almost eighty thousand accessions, and it is expected to add to the collection. Research at IRRI is also defensive, a constant search for new genetic sources of resistance—lest the old defenses break down or new pest biotypes appear. And it has trained thousands of young agronomists in rice production. Such work, humdrum as it is, has greatly benefited Asia's rice and Asia's farmers.

That IRRI has to keep in contact with farmers is surely true, particularly for production technology. The justification is both practical and ideological. Closing the yield gap has to be done in the field, not in the laboratory. Whether it is using more green manure and less fertilizer, pest management instead of pesticides, direct seeding instead of transplanting, or a new semidwarf instead of a low-yielding landrace, IRRI has to understand what works, what does not, and why. For this, links with on-farm projects are vital and irreplacable. On-farm work, however, is often the least valued. Extension lacks the prestige of scientific research. To downplay direct work with farmers only legitimizes a predisposition entrenched in much development thinking.

In the 1990s, environmental problems have upstaged old standbys like food production. And the list of problems is long. There is too much trash, toxic waste, and sewage. The earth's ozone layer has holes and the atmosphere is soaking up too much greenhouse gas. The planet's genetic inheritance is being squandered.

Almost all these problems effect agriculture. If the planet gets too much solar radiation, if the earth gets warmer, if our fields get contaminated by toxic waste, and if the world's biodiversity narrows, then the crops we depend on most will be endangered. The margin for error, the difference between self-sufficiency and hunger, is small. To feed a hungry planet under environmental stress is a great challenge. The work of IRRI is as vital today as ever.

2 BRAZIL

"Forget the centers," Ed Pulver told me, "what matters is local production." An American agronomist, Pulver had worked with the rice program at the International Center for Tropical Agriculture in Cali, Colombia. "Start in the field where the problems are, start with national programs."

When I finally tracked down Ed Pulver, he had left CIAT and moved to Carlyle, Illinois. It was the winter of 1989. "When you spend time in the field, the problems are obvious—they hit you over the head," he said. "Critics think cutting back on fertilizer should be a priority. Rice farmers don't think so. In Colombia and Brazil, fertilizer accounts for only 5 percent of production costs. So from the farmer's point of view, fertilizer just isn't a problem. But ask them about pesticides. Spraying fields is very expensive; it accounts for almost 25 percent of what a farmer spends to plant rice. What we've got is a pesticide problem. The priority should be pesticide use: how much is applied, when, and for what reasons." We talked into the night. When I left his makeshift house-trailer the next day, I revamped my plans. Instead of starting with rice at CIAT headquarters in Cali, I went first to local rice programs in Brazil and Chile, taking the list of names Ed Pulver gave me.

CONTRASTS

Potatoes are a marvelous crop of great diversity: from the red Pontiacs and yellow Inca golds to the Irish cobblers and fingerlike purplish Caribbeans, and of course the Russets, the Idahos, and the Kennebecs.

But rice is rice—just add water and boil—right?. How wrong that is. As I saw in the Philippines, rice comes in myriad shapes, sizes, and colors. South Asia prefers indica rices, East Asia the sticky japonicas. There are irrigated and rain-fed rices, floating rices, and upland rices. For monsoon Asia, rice is a way of life, a social system as much as it is a crop.[1] In the end, what rice is depends as much on cultural geography as on biology.

Rice is the main crop of monsoon Asia. To think of Asia is to think of rice. In Latin America, rice is mostly a cash crop rather than a locally traded subsistence crop. Latin American culture suggests many images, but a rice bowl is not one of them. Rice is so basic to Asia that its importance in Latin America is often overlooked. In volume, the region's production seems minor. But in Brazil and Colombia, in Ecuador and the Dominican Republic, per capita rice consumption exceeds 32 kilos, making it comparable in the diet of ordinary people to rice in Asia.[2]

Brazil is Latin America's largest rice producer. Per capita consumption of milled rice is 43 kilos a year. Compared to an average of 8 kilos in the United States, that seems like a lot, but it is low by Asian standards. Even in a moderate rice-eating country such as Japan, per capita consumption is 62 kilos.[3] Or consider the scale on which farmers produce rice. In Asia, most rice farmers get by with 1 hectare or less of land. Two hectares of rice is a big farm. Colombia's rural development agency, by contrast, defines as "small" any holding of less than 20 hectares. For rice in Colombia, 15 hectares is still small-scale production.[4]

DEVELOPMENTS

High-yielding indica semidwarfs from IRRI transformed Asian rice production. Founded in 1967, CIAT was modeled on IRRI. The rationale behind CIAT's rice program was that Asia's rice revolution could be transferred to Latin America with comparable results. Rice, however, is not CIAT's only job. It also works on the genetic improvement of beans and cassava (maniac) and looks for practical ways to restore degraded pastures and reduce erosion.

Much of CIAT's research is carried out with the small farmer in mind. In beans and cassava, it tries to improve yields without relying on chemical fertilizers or pesticides. With rice, however, modern inputs are essential to high productivity. And with ranching, small is a relative concept. Even 100 hectares is not much when cattle have to survive a long dry season. In the meantime, quite apart from any impact CIAT had, the nature of agriculture in the region has changed enormously.

Between 1960 and 1980, most Latin American governments adopted policies to modernize agriculture. The state expanded subsidized credit, promoted mechanization, improved the transportation system, and built storage facilities.[5] The overall results are impressive. During the 1970s, the increase in agricultural output kept ahead of the region's population growth. The area devoted to crops increased by 18 percent, and the harvested volume by more than 30 percent. Poultry production almost doubled, and the egg supply grew by half. Beef and pork output jumped by an average of 30 percent, and milk registered an increase of 37 percent.[6] The agricultural sector's reliance on modern inputs grew apace. When progress is tallied by the number of tractors used (up 42 percent), the quantity of fertilizer purchased (up 88 percent), and the volume of pesticides applied (up 75 percent), progress there was.[7] When who benefited from the changes is considered, qualifications are in order. Credit went primarily to those who had the collateral and know-how to use it profitably. Small farmers, even if they held title to their land, rarely benefited from state-backed subsidies and incentives. Lacking sufficient assets, unable to fill out forms, and caught in the clutches of assorted middlemen, they were the kind of high-risk, time-consuming clients the banks avoided. Big producers gained the most, small holders profited little, and sharecroppers got displaced. That being the case, it is not surprising families left the countryside. Between 1960 and 1980, the portion of the region's population that lived in rural areas declined from more than half to about a third.[8]

Brazil's transformation from a rural to an urban society occurred in just fifteen years. In 1965, half the country still lived in rural areas; in 1980, only 32 percent did. Brazil began the 1970s with four cities above the million population mark but ended it with eleven cities in that category. In just a decade, metropolitan São Paulo, the country's largest urban area, increased its population by more than 50 percent, from 8.1 million in 1970 to 12.6 million in 1980.[9] By comparison, the U.S. population, which struck a balance between the city and the countryside in 1920, took forty years to become as urban as Brazil.[10]

Sheer numbers alone make the point. Brazil is Latin America's most populous country. In 1965, it had an urban population of about forty million. Fifteen years later, the number of people living in cities had doubled to near eighty million.[11] Could the country provide decent housing, gainful employment, and urban services for the migrants who flooded into the cities? No. The hidden cost of Brazil's modernized agriculture is now paid out in abandoned street children, sacked supermarkets, and urban insecurity. Most of Latin America followed the pattern of rapid urbanization. Exceptions are Paraguay and Bolivia, which in 1993 still had about half their populations in rural areas, and Central America, where most countries were still at least 40 percent rural.[12]

Was the cost modernization imposed compensated for by increased production? The statistics lose much of their luster when scrutinized. Crop production is a good example. Export commodities, such as soybeans, sugarcane, and cocoa, did the best. Food crops, such as potatoes, maize, and beans, lagged far behind.[13] The reason was that many big producers favored exports or concentrated on the supermarkets and fast-food chains so popular with the middle class. Food processing in all its frozen and reconstituted forms became a vertically integrated growth industry. The result was prejudicial to the crops the majority of people depended on.[14] By the late 1970s, the region's food imports were increasing twice as fast as during the previous decade.[15]

Modernizing Latin America's agriculture turned out to be a great disappointment. The rich got subsidized credit, the poor got expelled from the countryside. Most countries ended up with an impoverished urban population with few prospects. That such a pattern is not an inherent aspect of modernization is amply demonstrated in Asia. Agricultural policy aimed to keep rural families on the land, not expel them to the cities. Had India followed Brazil's prescription for development, its urban population would have increased by more than two hundred million between 1965 and 1980—a staggering figure. It is hard to imagine that India could have sustained a change of such magnitude. Instead, India's urban population increased from 19 percent of the country's total in 1965 to 22 percent in 1980 and to 26 percent for 1993. And India is not unique. For Indonesia, the figures are 16 percent (1965), 20 percent (1980), and 33 percent (1993); and for Bangladesh, 6 percent (1965), 11 percent (1980), and 17 percent (1993).[16]

In Asia, most governments realized that modernizing agriculture had to be done with small farms and the rural labor force in mind. The opposite was true in Latin America. Nonetheless, the region still has a great stake in its small farms. At the beginning of the 1980s, they accounted for 80 percent of all holdings, although such farms added up to only 20 percent of the region's agricultural land. On this they grew half the region's maize, 60 percent of its potatoes, and 75 percent of its beans. Overall, small farmers produced 40 percent of the food destined for domestic markets, and they did so without the credit subsidies and expensive technology that agribusiness depended on.[17] Argentina and Chile aside, even in 1990, a third of Latin America's population still lived in rural areas.[18]

In Brazil, agriculture is still the country's single most important employer; in 1994, it provided jobs for 22 percent of the labor force.[19] Mechanization, however, has driven migrants to the city faster than the urban economy can employ them. Given the rapid population growth in cities, self-sufficiency in food crops is a matter of considerable urgency.

The task of CIAT is to conduct research and develop production technology for rice, beans, and cassava. In almost every Latin American country, at least one of these crops is a dietary staple. As to rice, CIAT was expected to adapt high-yielding indica varieties to local soils, diseases, and insect pests. And it did. To what extent did CIAT's work benefit the region's small producers? It depends on the country and specific rice zone in question. Between 1992 and 1994, Brazil accounted for more than half of Latin America's rice production. So it is good place to start.[20]

RICE IN BRAZIL

Compared to 1971–75, Brazil's total rice tonnage had risen by 43 percent for 1985–90: from an average of 7 to 10 million metric tons a year. Some of the gain came from an expansion in the country's rice acreage, which increased from an average of 4.8 to 5.4 million hectares. Thus when overall yields are considered, gains are rather modest; they rose from an average of 1.4 metric tons per hectare for 1971–75 to 1.9 metric tons for 1985–90.[21] By Asian standards, Brazil's performance could hardly be called a rice revolution. In fact, Brazil's population grew much faster than its rice production: from 93 million in 1970 to 150 million in 1990, a 50 percent increase.[22]

The reason Brazil's yield gains are modest is that 75 percent of its total rice acreage is in upland production. Only India produces as much upland rice as Brazil, but in India's case, upland acreage is only 15 percent of the country's total rice land. Recall that upland rice means direct seeding in a dry field—without irrigation or diked fields to collect rainfall. In an upland system, rice is produced dry like other cereal crops. Upland yields are notoriously low—often not more than a metric ton of rice per hectare.[23]

To find Brazil's rice revolution, we have to look where rice is irrigated. That is in the country's two southernmost states—Santa Catarina and Rio Grande do Sul. For 1990–93, they averaged a million hectares of rice, which accounted for only 22 percent of the country's rice land. On this they produced an average of 4.7 million metric tons of rice a year, almost half of Brazil's total.[24]

SANTA CATARINA

Brazil's territory covers more than 3 million square miles, making it the world's fifth largest country. With almost 156 million people, it is the sixth largest in population. Of Brazil's twenty-six states, Santa Catarina is one of the smallest. With some 37,000 square miles, it is larger than Ireland but ranks only twentieth in territory. In 1995, it had 5 million inhabitants and ranked eleventh in population, far behind Brazil's most populous state, São Paulo, which has 34 million. In food production, however, Santa Catarina ranks fifth, making it one of the country's most important agricultural states.

Santa Catarina's agroindustrial economy is the most balanced and diversified in Brazil. Light industries embrace a wide range of manufacturing activities, from food processing to textiles and machine tools; small firms predominate. In agriculture, small- and medium-sized family farms are the norm. Plantation agriculture, with its monopoly on land, slaves, and retainers, never dominated the state as in the sugarcane zone farther north. During the 1890s, when millions of immigrants from Italy, Germany, and Poland went to Brazil, they headed for southern states like Santa Catarina where labor was scarce and land abundant.[25]

Most of Santa Catarina's farmers plant several different cash crops. They also specialize in small-scale animal production, from milk to pork and poultry. The state's seven agricultural research stations reflect this diversity. They have experts in cereal crops, in vegetable production, and in fruits, both temperate and tropical. There are research stations for dairy farming, pasture improvement, and farming systems.

The state's geography helps explain its diversified farming output. The Atlantic coast, protected by mountains, has subtropical temperatures; in the winter, even light frost is rare. Along this coastal strip, rice is the most important crop, followed by cassava and citrus fruit. Across the mountains, a plateau spreads out through the central part of the state, making for a temperate zone of cool-weather crops like onions and wheat, apples and peaches. On the plateau, winter frost and snow are common. Finally, to the west, the state borders Argentina and the Uruguay River, which farther south empties into the Rio de la Plata estuary. In the river valleys of the west, the climate is even more tropical than along the coast.

At the Itajaí research station, Richard Bacha, an agronomist in his late forties, heads up the state's rice program.[26] He seemed relaxed in a pair of old jeans, boots, and a faded short-sleeved shirt, as much at home in a rice field as in an office. It was February and the middle of summer. Itajaí is an attractive port and textile center. The research station is a few miles out of town, in the heart of the state's rice belt. I spent several days there, visiting rice farms and experimental plots. Wherever I went, I heard the same story—what makes the state's agriculture so productive is a combination of good research and well-organized extension.

The extension service was set up in the mid-1950s, long before Brazil's federal government designed a system for the whole country. So Santa Catarina had a head start. Extension is based on the American model and was set up in consultation with U.S. agronomists. A serious undertaking by hardworking people with a well-defined approach, the extension service in Santa Catarina was soon considered the best in the country, and it still is. Later, when the federal government set up a crop-oriented research system to work with the states, Santa Catarina, in turn, set up its own research network to work with extension. The first research station for the Agrarian Research Agency of Santa Catarina (EMPASC) opened at Itajaí in 1975.[27]

"We had discipline and a sense of responsibility," Bacha said. Itajaí scientists worked jointly on training with extension personnel, and they helped set up demonstration plots, which were managed by local cooperatives. "We identified on-farm priorities first and then worked out a scheme that took them into account. We were demanding. When an approach did not work well, we rejected it." People were well-paid; they were encouraged to get advanced degrees in agronomy at the best universities. Consequently, both extension and research got off to a strong start.

Rice Report

Santa Catarina's traditional rices are tall and easily damaged by high winds. Compared to modern varieties, they have few panicles and set little seed. The Itajaí research program got underway at about the time indica semidwarf rices were developed. Although dense seed producers, the semidwarfs had some drawbacks. Harvesting on the state's small farms was still done manually; the shorter modern varieties made this a back-breaking task. And consumers did not like the cooking characteristics of the new rices, which tended to be sticky and thin. The rice industry resisted too. In Santa Catarina, rice is soaked in hot water, or "parboiled," before it is dried and packaged. This stabilizes the outer layer, which has most of the vitamins, and increases both the product's shelf life and weight.[28] Because the machinery and processing techniques in use best suited local varieties, the industry opposed changing its equipment and production methods.

"We did not impose modern varieties or new production technology on farmers," Bacha emphasized. Working with extension, Itajaí agronomists planted the semidwarfs next to the traditional ones, letting farmers observe the difference and decide what they wanted. The situation for rice was unique. According to Bacha, they had research that demonstrated the superiority of semidwarf varieties. They had excellent technical assistance available through extension, and new, small-scale harvesting equipment was developed. All the elements for success came together.

Improved varieties were so productive that pressure from farmers forced the processing industry to change over. Where farmers used to get 2 to 3 metric tons per hectare, they now got 4, 5, or 6. So a moderate drop in rice prices was offset by higher productivity. The new varieties quickly had a tremendous impact, and consumers came to accept—even prefer—the glutinous rice. Between 1975–76 and 1985–86, average yields rose from 2.2 to 3.2 metric tons per hectare. "If we subtract out the state's upland rice," Bacha said, "with yields of only a ton, the average for irrigated rice is closer to 4 tons per hectare." For 1991–93, the state harvested an average of 628,000 metric tons of rice from 142,000 hectares for an average yield of 4.4 metric tons.[29]

Once the new varieties were accepted, the main task was to improve them genetically so they tasted better and resisted local pests and stresses. New rices improved at Itajaí eventually replaced the original dwarf varieties. The state's *Technical Bulletin* for 1989 had yield data on ten modern rice varieties, half of which came from the breeding program at the Itajaí station.[30] Today, the priority is to develop an iron-tolerant variety, given the high iron content in coastal soils. On the list too are blast resistance and early maturity.

Santa Catarina's success in rice is due not just to its work in breeding. "New production practices are necessary with the semidwarfs," Bacha said, "which farmers have to apply to their rice fields." Farmers level and grade the land for better water management, they use nitrogen fertilizer, and prepare the soil better prior to planting. The Itajaí station also works on seed production and the direct seeding of rice paddies.

For a new variety to make headway, farmers need ample seed stocks on hand at planting time. Multiplying seed quickly, without a loss in quality, is a key aspect of the state's rice revolution. Depending on density, it takes at least 115 kilos of seed to plant 1 hectare of rice. Even though Santa Catarina has only about 140,000 hectares in rice, that still adds up to millions of kilos. Of course, many farmers hold over seed from their last crop and grow several varieties. Nonetheless, if not worked out in advance, inadequate seed production prevents a new variety's rapid spread. In Santa Catarina, the Itajaí station and the extension service select the best farmers in each district to multiply seed. They provide each with a sack of seed to start multiplication—enough to produce about twenty-five sacks, distributed as follows: eleven sacks are sold locally, ten sacks go to the grower, and four sacks go back to extension for distribution and subsequent multiplication by other farmers. In this way, a new variety spreads through the rice zone in just two seasons, so quickly that the loss in seed quality is minimal.

For direct seeding, the Itajaí station favors pregermination.[31] Seed is soaked for between twenty-four and forty-eight hours, which increases its weight by about 25 percent. The soaked seed is then sown in a shallow flooded field about 50 millimeters deep. The heavy seed falls though the water to rest in the mud below. Three days later, farmers let the water level drop so that normal plant growth can occur in saturated soil. Once the rice has sprouted and is growing, the field is flooded again.

Bacha took me to experimental plots near the Itajaí station. He has developed a simple method to identify iron-tolerant varieties. He plants the test varieties between a resistant and a susceptible strain: anyone walking by the trial site can compare the results visually. The station also has trials underway for resistance to blast, a fungus disease that dries up the rice plant and reduces yields. Many of the varieties tested come from IRRI and CIAT. In some plots, azolla was being tested for nitrogen fixation, but it does poorly in Santa Catarina's cooler climate.

Farmers plant rice in the spring, which in Brazil is from late September until early December (seasons in the Southern Hemisphere being the reverse of North America's). The harvest comes between 120 and 150 days later, depending on how early the varieties planted mature. When farmers plant several hectares of rice, they usually stagger planting dates to spread out the harvest. That is important, because few farmers have drying and storage facilities. They sell almost immediately, which usually means lower prices. "Investing in storage capacity pays off in the first year," Bacha noted, "but farmers cannot get the banks to loan them the money they need, not without years of lobbying state agencies."

Santa Catarina's farmers have a reputation for the quality and diversity of their output. Nonetheless, it is tough to be a farmer. "Young people want to get jobs in local textile mills and in industries," Bacha said. "The work is easier and more secure; they get holidays, medical care, and retirement. Farmers cannot afford to get sick, much less retire. Most of our rice farmers are middle-aged, in their forties at least. I doubt that their children will take over."

RIO GRANDE DO SUL

Brazil's southernmost state, Rio Grande do Sul is also the country's coldest. Freezing temperatures are common in winter, even along the coast; in the mountains there are ski resorts. Strong winds and cold fronts come up from Argentina farther south. So too does much of the state's geology. The great Argentine pampas sweep north across Uruguay into the south-central part of Rio Grande do Sul. The state's distinctive character comes more from ranching and gauchos than from rice.

A middle-sized state for Brazil, Rio Grande do Sul is twice the size of New York state. On the northern plateau, a region famous for its wheat and soybeans, temperatures are cool, even in summer. In the mountains that separate the plateau from the coast are vineyards, which produce the country's best wine. Rice is grown in the southern half of the state; almost all of it is irrigated. During 1990–93, Rio Grande do Sul produced 45 percent of Brazil's rice, an average of 4.1 million metric tons a year on 849,000 hectares.[32]

The rice zone's topography is diverse. Wedged between the pampas and the plateau is the state's central basin, drained by the Jacuí River and its tributaries. This is the state's oldest zone of production, an area of small family farms. In counties such as Cachoiera do Sul, Agudo, and Dona Francisca, more than 20 hectares of rice is a lot. Yields are high, between 5.2 and 6 tons per hectare.[33]

During the rainy season, farmers dam up small streams or collect the runoff in natural depressions. The water so stored is gravity fed into rice paddies during the summer months. To get water directly from the rivers usually requires an expensive investment in pumps and in energy.

Just to the north of Porto Alegre, the state's capital, the mountains cut down across the coast and out into the sea. To the south is a broad plain formed by the largest inland bay in South America, the Lagoa dos Patos. The land in this Southern Region is the flattest in the state, ideal for mechanization. In counties such as Pelotas and Capão de Leão, the typical producer has at least 50 hectares in rice. And farther south still, between Lakes Mirim and Mangueira, many agribusiness growers have 300 to 500 hectares in rice, some more than 1,000 hectares.[34]

With easy access to water and a terrain of gently sloping land, the Southern Region has low irrigation costs. A shorter growing season, however, offsets this advantage. In the central basin, for example, most farmers have their rice fields planted by mid-November; for the South, it is the end of the month or into December.[35]

In the heart of the state's cattle country, southwest of Porto Alegre, the Campanha Region is also a big rice area. Small dams provide irrigation, most of them located near geologically natural sites such as riverbeds, watersheds, and depressions that fill up during the rainy season. A special federal program had funded this small-scale approach to local irrigation, and many farmers in the Campanha have benefited. Before it was phased out in the late 1980s, the program claimed it had helped irrigate more than 1 million hectares nationwide. The Campanha Region's chief advantages are its good soils, which were only recently brought into production, it's dry summers, and it's high luminosity. The main limitation is insufficient water.

The newest rice zone is in the Western Region, by tributaries that feed the Uruguay River. Soils are rich and alluvial, but erosion takes its toll. The region's fertile river valleys tend to be hilly or undulating. So water often has to be pumped up to holding tanks at different elevations and then gravity fed into the fields below. This can erode the soil, particularly when fields are not adequately leveled. Nonetheless, counties in the Western Region, such as Itaquí and Uruguaiana, have taken over as the state's top rice producers.

Santa Catarina is a bastion of small family farms whose families own the land they work. Most of the state's rice production is on plots of 10 hectares or less. In Rio Grande do Sul, by contrast, two-thirds of the rice is grown on rented land, and it is much harder to characterize the typical rice farm. In 1989, the average grower had 70 hectares in rice. However, almost five thousand rice farmers—half the state's total—with 25 hectares or less jointly held a scant 6 percent of the crop land devoted to rice. Some 250 farms with more than 400 hectares apiece collectively accounted for a quarter of the state's rice land.[36] Santa Catarina has a homogeneous rice zone. In Rio

Grande do Sul, by contrast, rice zones differ by farm size, by climate, and by production problems.

The Rice Agency

When I visited Rio Grande do Sul in 1989, Dr. Paulo Carmona was in charge of genetic improvement at the state's Rice Research Agency (IRGA).[37] He had a patience, a presence of mind, and a capacity for getting things done that drew people to him. Carmona had twenty years of experience and was devoted to the state's farmers, to agriculture, and to rice. At the International Rice Conference held in Mexico City in 1991, he received a Distinguished Rice Scientist award.[38] The research station at Cachoeirinha is just north of Porto Alegre; Carmona showed me around and explained its work. Later, we went to trial sites in three of the state's principle rice zones.[39]

Rio Grande do Sul's IRGA receives funds directly from a tax levied on each sack of rice processed in the state. The tax made the agency self-supporting and relatively autonomous. Nonetheless, it often got short-changed. The state government diverted part of the agency's funds or imposed unwarranted restrictions. In 1989, the travel allowance for field trips was so low that researchers paid most costs out of their own pockets. So while IRGA had its own funding, in practice, state politics interfered with its work.

Rio Grande do Sul's rice regions are subdivided into thirty-seven county-based extension zones, and IRGA has an office in each zone with at least an agronomist and an assistant. Carmona considers agricultural extension the agency's main job. Big producers are usually aggressive adopters of new technology; they can read up on research results published in IRGA bulletins. It is the smaller farmers who need extension. The agency's research program is closely linked to on-farm problems. Field trials, for example, are always done jointly with the extension service. And each year the agency's research division presents its agenda to the extension staff for criticism. "Compared to CIAT or Brazil's National Rice Center in Goiânia," Carmona said, "we have much more direct contact with on-farm problems."

Rice Production

In 1900, Rio Grande do Sul was already Brazil's principal rice producer. Even though many farmers irrigated their rice crop, yields were low and improvement gradual.[40] Between 1921–25 and 1961–65, yields in the state rose from an average of 2.1 metric tons per hectare to about 2.5 tons.[41]

Farmers planted traditional, tall varieties, rarely used fertilizer, and did little to level the terrain. Significant gains started in the late 1960s, a consequence of improved soil preparation, herbicides, and careful crop management.

Rio Grande do Sul's climate is more akin to that of Louisiana, a leading U.S. rice-producing state, than a tropical country like Colombia. So it is not surprising U.S. varieties including Louisiana's bluebelle, did well there. A pre-IR8 improved variety, bluebelle had relatively high yields and, according to Paulo Carmona, excellent grain quality.[42] But it was not a semidwarf rice. As bluebelle spread, yields in the state went up. In 1968, they reached an average of 3 metric tons and did not fall below that mark again. Subsequently, the state's rice acreage expanded rapidly.[43] Then, in the 1980s, the new semidwarf varieties came along—and just in time. "The cost of production," Carmona said, "was going up and higher productivity was needed to cover them."

In 1970, modern varieties were already well established in Asia. Why did it take another decade for them to reach Rio Grande do Sul? Part of the problem was grain quality. The first semidwarf varieties were too glutinous for local tastes, and farmers already had a quality rice in bluebelle. Climatic factors, however, were decisive. Suited to a tropical climate, Asian indica rices are more sensitive to the state's cold weather than a variety like bluebelle. Carmona said it took several years to adapt semidwarf rices to the state's temperate conditions. "We released the first modern variety in 1980," he said. "Its grain quality was high and in yields it clearly surpassed bluebelle." To show farmers the advantages of the new varieties, IRGA set up demonstration plots. Even though it knew they were better, it did not push the new varieties. It let farmers try them out on a limited basis first. Nonetheless, the semidwarfs spread quickly; today, they cover 90 percent of the state's rice area.

Cultivation also changed. Until bluebelle came along, harvesting by hand predominated. Since then, mechanization has taken over. Today, almost all farmers use fertilizer and apply a herbicide at least once. Farmers also pay much more attention to soil preparation and water control. Almost everybody uses or rents a tractor. Yields in Rio Grande do Sul average between 4.5 and 5 metric tons per hectare. The potential of the new varieties is about 7.5 tons per hectare; some counties in the state already average 6 tons.[44]

Red Rice

"The greatest obstacle to higher yields," Maurício Fisher explained, "is intrusive, weedy, red rice."[45] A tall, thin man in jeans and a t-shirt, Maurício works on weed control at IRGA. "In older regions of production," he said, "particularly in the Central Basin and the South, losses to red rice reduce yields by 20 percent—that's a ton a hectare."

In most of Rio Grande do Sul, farmers directly seed their rice crop when the fields are still dry. They use dry rather than pregerminated seed. After the rice sprouts and the plant has grown enough, they flood the fields.[46] The problem is that seeds from grassy weeds get churned up during the soil's preparation and germinate along with the rice. And in early stages of growth it is hard to tell the difference between an intrusive grass and the rice. Red rice is classified as *Oryza sativa* and is not really a weed. In the field, however, natural hybridization occurs between red rice and commercial varieties.[47] Farmers end up with hardy, red rice–type "weeds," all of which have undesirable, semiwild traits. Once a field is seeded, a herbicide that kills off red rice is also lethal to commercial varieties.

Red rice stands out when threshed because of its reddish tinge. But the problem is not really its color. It shatters much too easily. According to Fisher, red rice matures earlier than popular commercial varieties; almost 80 percent of it falls to the ground before harvest time. The red rice seeds get mixed into the soil, resprouting with the next rice crop.

To reduce red rice intrusion, the first step is to keep it out of the seed stocks. Certifying rice seed for sale, IRGA set quality standards high: at first, the allowance was no more than three grains of red rice in 500 grams of rice seed, and finally there was zero tolerance.

Most corrective measures depend on farmers. Because red rice is taller than commercial semidwarfs, farmers can apply a herbicide selectively to the tallest plants or cut off the top of the red rice before it seeds. Such labor-intensive measures are appropriate for a small rice farm but not for large-scale production.

In Rio Grande do Sul, rice production and cattle ranching are usually related activities. When farmers rest their rice fields, they allow a rustic pasture to take its place. After a field is drained in the fall, farmers simply leave it to revegetate. Rice grass comes up from the stubble left behind or from seeds that fall to the ground at harvest, including red rice. During the first year, rice grasses predominate, which the cattle like. The next year, the rice percentage falls as native grasses recover and start to take over. In the third year, farmers usually rotate back to rice. The rice-pasture rotation helps to restore the soil's fertility and control red rice.[48] During the rice-pasture cycle, farmers can take action against red rice. "Ninety days after the harvest," explained Fisher, "when the red rice resprouts, they can apply a herbicide. Also, the summer before they rotate back to rice, they can plow up the soil, which airs it out and brings red rice seed to the surface. After it germinates, they apply a herbicide."

When regular rice gets a headstart, red rice intrusion can be much reduced. The strategy here is to shift to pregerminated seed, as is common in Santa Catarina. In this way, the regular rice matures first and harvesting occurs before the red rice shatters.

Plant Improvement

A division of labor prevails in breeding. Research at CIAT develops new rice lines with desirable traits and adaptability to a range of rice environments. Breeders such as Paulo Carmona can draw on the CIAT lines for local testing. But this is just a first step: CIAT lines have to be crossed with cold-tolerant varieties that meet Rio Grande do Sul's special requirements.

Reducing red rice growth is a case in point. Most solutions add steps to production and increase costs. An early maturing variety, by contrast, can be harvested before the red rice matures; no special measures are needed. Early maturity has other benefits too. Heavy rains in the spring often delay planting. Then in the fall, if there are early cold spells and frost, yields at harvest time drop sharply. An early variety can escape such situations, a particularly desirable trait where there is a short growing season and less leeway to get the crop in and get it out. "Rice production can be a risky business," Carmona said. "Farmers have to maintain their irrigation systems, they have to pay for tractors and equipment, and for fertilizer and herbicides. They cannot afford to sacrifice high yields for a secondary characteristic like earliness. We have many early varieties, the problem is to associate high productivity with earliness in a region where cold tolerance is important."

The state's most popular rice varieties take 140 days to mature—compared to 136 days for red rice. The "semiearly" variety that IRGA was testing in different production zones matured in 125 days: a two-week advantage. That was just enough to beat out red rice and avoid early frosts.

The first semidwarfs held up well against blast disease, but the resistance is breaking down. And the most popular varieties are still sensitive to iron toxicity. So a basic objective is to associate blast resistance and iron tolerance without losing high productivity or earliness. Such modifications would keep yields high and reduce losses to disease. As Carmona explained, "When disease builds up, productivity falls. To maintain yields, we need many alternatives to older varieties. Producers want the variety that best fits their situation. Compared to the dramatic impact the first semidwarfs had, genetic improvements today payoff in smaller, incremental gains."

Evaluation

Before IRGA releases a new variety, it conducts trials in each of the state's rice zones, the last step prior to on-farm demonstration plots. I spent two days accompanying Paulo Carmona on his rounds to three testing sites—in the central basin, in the Campanha Region, and to the west.[49] It was mid-February, far enough along in the growing season to evaluate how well new lines had fared under field conditions.

The trial plots at Capané in the heart of the central basin were laid out in neat rectangles. I had brought rubber boots that came up to my knees. They quickly filled up with water and stuck in the mud, making it difficult to maneuvre without losing balance. I soon abandoned them and went barefoot like Carmona.

Most of the tests were for what Carmona described as "moderately early" varieties: 120 to 130 days. What he considered an "early" variety matured in less than 120 days. What made the site different was its poor soils. It is difficult to get high yields from early varieties because they have less time to set grain and ripen, and because their yields are particularly sensitive to poor soils. So a variety that does well here ought to fare even better elsewhere. After examining the trials, Carmona felt he had a promising line that was moderately early (125 days), reasonably productive, and iron tolerant.

Rolling hills and mountains separate the central basin from the Campanha Region to the south. The Campanha is noted as much for its ranching as for its rice. Located in the northern reaches of the pampas, it has hot, dry summers. Irrigation comes from small dams and ponds. But in 1989, the rains stopped early and a drought set in. Shallow reservoirs and marshlands had already dried up. To cut their losses, many farmers reduced the area irrigated. Yields were expected to drop by a third.

We headed for Bagé, only 70 miles from the Uruguay border. It is the zone's largest city, and IRGA has its regional headquarters there. In Rio Grande do Sul, no social exchange is complete without *chimarrão*, a tea made from maté. Boiling water is poured over the maté leaves, which have been crushed in a bottle gourd and left to steep. The gourd is passed around and the maté sipped from a stainless steel straw. So too at the Bagé office. When the gourd is empty, more hot water is added and the ritual is repeated.

The Campanha trials were conducted near the campus of the region's agronomy school. The soil is fertile, composed of sandy clay good for irrigation. It was a hot summer day when I visited. The sky was deep blue, the sun had a brilliant, penetrating intensity. The air itself felt clean, bright, and clear—conditions described as "high luminosity." This means more than simply the sheer number of daylight hours, it means the clarity, the sharpness, the strength of the sunlight received—qualities difficult to measure but easy to feel. The Campanha is noted for its luminosity, which makes the region good for cereal crops, and for its cool nights, which help the rice plant conserve energy. "When nights are too hot," Carmona said, "the plant loses energy. The ideal is high luminosity combined with somewhat cool nights." In the Campanha, rice and vineyards were especially popular. California has high luminosity too, as does the coast of Peru—and in both cases, vineyards and rice production are well established.

The Campanha trials were inconclusive. The faculty member in charge had changed jobs, so the trials languished—overcome by weeds and starved for water. "We conduct trails at five different sites every year," Carmona said, "and we always lose a couple. At Capané, the rice agency oversees the work, so it is usually good. And we get good results with trials done by farmers. Nonetheless, a farmer does not see all the angles on a variety that a researcher does. We have done trials for twenty years and we still do not have a dependable system."

Bagé is a city of traditions. At night, families stroll along the main street, stopping for coffee and pastry at the town's cafés. To cut down on expenses, Carmona and I shared a room that night at the City Hotel. The next morning we headed northwest to Rosário, but not before a breakfast of cold cuts, assorted fruit, rolls, and lots of coffee with hot milk. The traditional breakfast in Bagé did not include rice.

Most of Rosário's rice production is on level ground between the Ibicuí and Santa Maria Rivers. A western area of recent production and high yields, red rice and weed problems are minimal. The Rosário office of IRGA is neat and well ordered. There are maps of every depression, marsh, and watershed in the district—as well as the associated dams and ponds. There is also a laboratory unit for sampling and analyzing the seed brought in for certification.

The field trials near Rosário were in excellent condition. The soils are light and sandy, but not excessively so; consequently, water retention is good without excessive absorption. The reservoir is higher than surrounding fields, so the water can be gravity fed into the paddies below. Large plastic tubes about 12 centimeters wide connect paddies. Inserted through the top of each dike, water passes from paddy to paddy—from one just a bit higher to one just a bit lower. Each trial is clearly numbered, with the key kept in a large notebook. Carmona waded though the paddies, making his evaluations visually. The moderately early variety that had impressed him at Capané did so here too. It has plentiful seed-bearing stems that set large, compact grains. He decided to recommend this variety to the extension service for on-farm trials next year.

Seed Production

Carmona and I returned to Porto Alegre that evening. As arranged, I headed to Pelotas, the largest town in the Southern Region, near Lake Mirim. The drought had effected the South too, despite its inland lakes. "As Lake Mirim's level falls, farmers have to extend their intake pipes farther out," Fernando Bruno said, "which increases production costs. Fortunately, Mirim is a freshwater lake. But brackish Lagoa dos Patos bay drains into the Atlantic. This year, with the drought, the water's saltiness is more pronounced."

A hefty, muscular man, Bruno had thirty years of experience in rice production and extension.[50] His job at IRGA's Pelotas extension office is to supervise seed production. Unless enough seed is produced for distribution, releasing a new variety is mostly a technical action with little impact. As an inspector, Bruno regularly checks the district's 26 seed producers on behalf of Brazil's Ministry of Agriculture, which has subcontracted the responsibility in Rio Grande do Sul to the state's rice agency. In the state overall, some 136 growers were certified to produce seed.

During field trips, Bruno makes sure growers take the steps necessary to produce high-quality seed. Later, when the seed is brought to the Pelotas office for certification, technicians take a sample from each sack. They check and record the seeds' germination rate, size, quality, and health, making sure it is disease free. The records are kept on file. If problems occur later in the field, they go back to double-check the tests to make sure the fault is not in the seed itself. Growers are careful to follow the guidelines set; otherwise, the agency can refuse to buy or to release their seed.

Bruno owns 250 hectares, inherited from his father, but he does not bother with rice. Instead, he rents the land for ranching. "The biggest problem in agriculture," he said, "is that Brazil has no coherent policy. Prices for rice are frozen, but the cost of everything else, including inputs, keeps going up. My land has so much red rice, I cannot afford to get rid of it."

Impatient with politics, Bruno feels that extension has suffered from shifts in policy. "We have a universe of producers," he said. "To reach them we have cooperatives and extension. But from our hypothetical 100 percent coverage we have to subtract out the counties without offices. And we can subtract the big producers who can seek out new technology on their own. That leaves a target population of small- and medium-sized farms. To reach them we do not have enough people, we do not have enough resources, we do not have enough equipment, and we do not provide enough training. The farmers reached are but a small minority."

Rotations

"With our research focused so much on rice," Paulo Carmona had said, "it is easy to overlook the farming system as a whole. Success cannot be measured by a single crop." In fact, diversification is an important item on IRGA's research agenda. "Rotating to a different crop," Carmona said, "reduces disease problems, helps the soil to recuperate, and aerates it too, compared to grazing, which compacts it. As a result, rice does better after a rotation."

Brazil's National Center for the Temperate Lowlands (CPATB) also works on rotations.[51] Because the center is located near Pelotas, Carmona suggested a stop there. Most of Brazil's national research centers have crop specialties, but in a few cases, as at Pelotas, they work more on production environments. I spent the day at the center discussing rotations.

The common rotation in the state is from rice to pastures. Some 80 percent of the rice acreage is recycled into a grazing system.[52] Most farmers plant rice for two years in succession, then let the field return to pasture for three, even four, years. During the recuperation period, they rarely apply fertilizer or herbicides to the field. In terms of chemical buildup, therefore, the rotation does not add to the region's ecological problems. Nonetheless, a low-grade pasture results in less milk production, less weight gain for cattle, and more grazing land per head. At CPATB the objective is to improve pasture productivity.

The land stays in pasture for three to four years," said Dr. José Carlos Reis, "so it is worth establishing a good one.[53] Milk production and weight will not go up on rice stubble or native grasses, which stop growing in the winter." Reis headed the center's research on pastures. The most popular improvement so far has been to plant rye as a winter forage. Still better, however, is seeding with a high-quality grass such as fescue. Even when farmers add a crop like soybeans to the rotation sequence, they still end up with the land in pasture for two or three years. Considering the cycle overall, a better pasture is a key component.

Adding a new crop to the rotation system can be difficult. Sorghum is attractive to agronomists because it is so drought tolerant—farmers do not have to divert water from their rice paddies to get good sorghum yields. It also makes for a high-quality animal feed. But sorghum is still a novelty in Brazil, and as a novelty, its commercialization has yet to be successfully worked out, making it too risky for most farmers. Soybeans, by contrast, are popular and lucrative. With production averaging 54 million metric tons for 1991–93, the United States is the world's top soybean producer. Brazil ranks a respectable second with its 1991–93 production at 19 million tons a year. The country's vegetable shortening and cooking oil come almost exclusively from soybeans, and Brazil has a thriving export trade in semiprocessed soybean products.

Although they are not as drought tolerant as sorghum, soybeans do well in Rio Grande do Sul. Good drainage is important, as the crop dislikes saturated soils. So, after the rice has been harvested, the fields have to be regraded for drainage during the rainy season. Soybeans need a rhizobial inoculant to stimulate nodule growth on the plant's roots and hence nitrogen fixation. With good nodule growth, the right soil conditions, and favorable weather, a soybean crop fixes so much nitrogen that plenty is left for a subsequent rice crop.[54] About 10 percent of the state's rice land is rotated to soybeans.

Francisco Vernetti specialized in soybean production at CPATB.[55] "Two crops of soybeans planted prior to another rice crop virtually eliminate red rice," he said. "That's not so true with a pasture rotation. Rice is a grass, so a rice to pasture rotation is really a grass to grass rotation; the weed problems just get worse." By contrast, the root structure and nutrient requirements of the leguminous soybean differ greatly from a grass like rice. Soybeans add nitrogen to the soil, which helps increase subsequent rice yields, and it is a good cash crop.

In the lowlands, most farms have at least 100 hectares. So they invest in farming equipment, tractors, and storage facilities. With a second crop, particularly one that fetches as good a price as soybeans, the cost of the equipment relative to what is produced falls. This benefit holds up even with a smaller cost-benefit margin, as is true with winter wheat.

"Wheat yields are poor in the lowlands," Vanderlie da Rosa Caetano admitted. A strong-willed man, Caetano brings great determination to his work on wheat. His specialty is plant pathology, particularly virus diseases. The region's climatic conditions are, in fact, adverse for wheat. During the winter, luminosity is low, with half the days cloudy. Then come strong winds and heavy winter rains that soak the soil, followed by high humidity. Little research is done on wheat varieties for such marginal conditions. In Mexico, the International Maize and Wheat Improvement Center (CIMMYT) focuses on more promising environments.

Although the lowlands do not have a good climate for wheat, the region's farmers have no other options for the winter. Because most of the equipment used to grow rice can be adapted to the requirements of wheat, farmers do not need high profit margins to justify the crop. Brazil's population is growing, and so too is the demand for wheat.

According to Caetano, wheat yields on CPATB's trial plots have reached 3.8 metric tons, which is excellent by Brazilian standards. When planted by farmers after a rice crop, the yield dropped to 1.9 tons—still sufficient to cover costs and justify production. Nonetheless, a yield potential of 5 to 6 tons is the objective. That left more room for the many pitfalls production has to face. In Caetano's opinion, the world's population cannot be fed unless crops like wheat can do well in marginal settings.

Administration

The administrative headquarters of IRGA were in Porto Alegre, Rio Grande do Sul's state capital. I walked to the office from my hotel. It was outside the city's old quarter on a broad boulevard lined with trees and flowers. Paulo Carmona arranged for me to meet the agency's director, Angelo Soares. A busy, contentious man in his late forties, Soares quickly went over the agency's history, its current priorities, and the geography of the state's rice production.[56] Rio Grande do Sul could greatly expand its rice acreage, he said, particularly in the South where water is abundant. The state already has the agroindustrial base for this: good transportation, hydroelectric power, a substantial storage capacity, and more than fifty mills that process fifteen million sacks of rice annually, most of it parboiled. Soares argued that the main obstacle, for rice and for Brazil's agriculture in general, is political:

> The mortal sin in agriculture is overcentralization. The federal government in Brasília sets the policy for agriculture; that will never be suc-

cessful. In agriculture, we have to interrelate short-term, medium-, and long-term work. But a centralized approach does not provide the right conditions for this. Every time there is a shift in leadership at the Ministry of Agriculture, priorities and policies change almost immediately; there is no continuity.

Policy has to be set at the local and state level; and you need a distinct policy for different crops. Agriculture is the administration of possibility, not certainty. Floods, drought, hail, credit problems always intervene: but in different ways, in different places, and at different times. Brasília can define general guidelines, but the specific application has to be done in a flexible fashion locally.

Consider the tremendous differences in soil quality, rainfall, climate, socioeconomic conditions, and ecosystem characteristics that underlie the state's rice production. Agriculture is regional and highly susceptible to local variation. How different this is from industry. Build an automobile factory in Siberia and it will still produce cars. Plant a sack of rice seed and the harsh climate will destroy the crop. Brazil exports shoes, good for feet here, good for feet in the United States. Exporting a meal, by contrast, is much harder; local tastes and customs intervene. In Brazil, we put rice and meat on the same plate. In Uruguay, they come on separate plates. And in the United States, meat usually comes with a baked potato. When Americans serve rice, to my taste it is simply awful. So in agriculture we have to work with a different mentality because it is less cosmopolitan and more local.

A WALK

Porto Alegre's old quarter is a maze of winding streets sheltered from the automobile. On a long weekend in a strange city the mind wanders through its own narrow passages, turns in upon itself. I walked through old, dusty shops: hardware stores where merchandise lined the walls, tailor shops where suits still came forth from whole cloth, and the city's central market, smelling of people, fish, and produce. It is a world besieged by shopping malls, supermarkets, and fast-food chains, a world undone long before we recognized its worth. Over *cafezinho*—a demitasse of steaming hot coffee, deep, dark, and sweet—I thought about how our age is addicted to the elegance of its concepts. We like our problems analyzed thoroughly, but we do not have the patience to bother with solutions.

THE CERRADO

Brazil's interior, its vast *cerrado*, is comprised of great plains and mesas. The region covers a quarter of Brazil's territory and is twice the size of Texas. Until the 1960s, it was a sparsely populated frontier of forests, scrub brush, and scrawny cattle. When the country's capital was transferred from the coast to the cerrado in 1960 (from Rio de Janeiro to the new city of Brasília), the pace of internal migration and settlement accelerated. Ranching, rice, and soybeans replaced the cerrado's scrub brush and forests. As the region's development intensified, the old states of Mato Grosso and Goiás were subdivided, creating two new ones: Mato Grosso do Sul and Tocantins.

During Brazil's summer, the moist air of the Amazon sweeps south and drenches the region in rain. During the winter, the arid antarctic winds prevail: the rains end, the earth dries up, dormancy sets in. To offset winter's dangerously low humidity, planners dammed up a river to make a huge lake on Brasília's outskirts. Nonetheless, when the dehydration index gets high enough, schools are closed and the citizenry told to drink water and take salt tablets.

When I visited Brazil's Agricultural Research Center for the Cerrado (CPAC) in 1988, it was August and a dehydration warning was in effect.[57] The center researches crops suitable to the region's conditions and ways to improve pastures. Distinct ecologically from the Amazon basin to the north, the cerrado can be cold in winter, has a prolonged dry season, and many of its rivers flow southwest into the Paraná-Paraguay system or northeast into the São Francisco River. Nevertheless, opening the cerrado brought farming, ranching, and mining to the doorstep of the Amazon rain forest.

I met with Dr. Carlos Magno Campos da Rocha, CPAC's director. He is an articulate, enthusiastic man with many years of experience in the region.[58] "Population pressure on Brazil's coast," Rocha said, "made the cerrado's settlement tremendously important. Consider the cost of land. Five hundred hectares in the cerrado cost less than 50 in Rio Grande do Sul. And climatically, the region has great promise for agriculture." Given its low tropical latitude, day length departs little from a constant twelve hours daily. Rainfall averages between 1,200 and 1,800 millimeters, and there is never a frost.

The cerrado has other advantages too. A topography of broad savanna plains with gently inclined slopes facilitates mechanized agriculture and road construction, and rivers and abundant rainfall make hydroelectric plants relatively cheap. By the late 1970s, the cerrado had all the electrical energy it needed, plus a network of paved roads, which meant market access for its agricultural products.

In 1975 the cerrado produced almost no soybeans. A decade later, output was close to 6 million metric tons, roughly a third of Brazil's total soybean harvest. In 1970, the cerrado had 20 percent of Brazil's cattle, but little by way of milk production. In 1989, it accounted for a third of the cattle—some fifty million head—and for 12 percent of Brazil's milk production.[59]

Cattle and soybeans are closely linked to rice. Before converting cleared land to pasture, farmers plant upland rice as a low-input cash crop. In 1970, the cerrado produced 1.8 million metric tons of upland rice, almost 25 percent of Brazil's total rice production.[60] With so much cheap land, low yields mattered less. When planted in successive seasons, however, upland rice depletes the cerrado's poor soils. Soybeans provide a nitrogen-enhancing cash crop that can be worked into the rice-pasture cycle. Nonetheless, despite Rocha's enthusiasm, agriculture in the region has its difficulties.

The cerrado's savanna soils are deficient in phosphorus and of very low fertility. They are so highly acidic that farmers must add lime to lower the pH level. Fortunately, the cerrado's rock phosphates and lime deposits are abundant, easy to get at, and cheap. Nonetheless, improving the soil and preventing erosion requires good farm management. Given the intense downpours in the rainy season, mechanization and poor contours can make erosion worse. And farmers migrating to the cerrado brought crop diseases, insect pests, and inappropriate agronomic practices with them. Making the cerrado productive is thus a tremendous challenge.

"Twenty-five years ago," Rocha said, "we knew very little about how to adapt crops to the cerrado's soils and climate. Now we do. During the past decade, agricultural production has grown much faster than the area under cultivation, which is to say that productivity has increased." The cerrado is a leader in rice and soybeans and is also diversifying into corn, coffee, and citrus fruit. The yield gap between the cerrado and other regions has narrowed. And with irrigation, the cerrado's yields are comparable to those elsewhere in Brazil. With respect to basic research and new production technology, in Rocha's view, Brazil is the leading country in the tropics.[61]

When I went to CPAC, I was interested in its ties to CIAT and the technology the two centers had for sustainable pastures. Because of the cerrado's long winter dry season, maintaining a permanent pasture is difficult. In Goiás, it stops raining in May and does not rain again until October. The problem is to find grasses and forage legumes hardy enough to survive. On low-grade pastures it takes 5 hectares to maintain just one head of cattle. By improving pastures, less land is needed, which means less pressure to clear new land in fragile environments.

Given the importance of upland rice in the cerrado, it is not surprising the country's center for rice research is located there. To promote the region's settlement, the federal government promoted a strong agricultural base as a first step. The CPAC is near Brasília, and the rice center (CNPAF) is just 200 miles away, near Goiânia, the state capital of Goiás.[62] The trip from Brasília to Goiânia is an easy four-hour bus ride on a well-paved highway.

When I went to CPAC, it was August and winter. The next year, when I took the bus to Goiânia and the rice center it was March and summer. I could not believe the difference. The parched brown fields and dust of August had given way to a humid tropical lushness. The miracle of rain brought forth a thousand shades of green in fields of beans and rice, of sugarcane and soybeans. Flatlands and mesas stretched out to the horizon. As on the Great Plains of the United States, sky and earth met in the distance. Given the cerrado's landscape, I was not surprised to find upland rice. What I did not fully appreciate then was how different upland and irrigated rice are.

Rice in the Cerrado

During 1985–90, Brazil had an average of more than 3 million hectares in upland rice.[63] In fact, half the country's rice, some 5 million metric tons a year, was grown as a dry, upland crop.[64] Within Brazil, the cerrado accounted for more than half of the upland rice total, and within the cerrado, Goiás was the leading producer.[65]

With upland rice, there are no dikes or elaborate drainage systems. Fields are plowed and seeded while dry. Farmers rely on upland varieties because modern irrigated ones need saturated soils to do well. The yield from upland rices, however, is notoriously low, and the productivity gap only got worse as irrigated rice improved. In 1965, for example, yields from irrigated rice fields in Rio Grande do Sul averaged about 2.5 metric tons per hectare; in Goiás, upland yields averaged about 1 ton. Twenty years later, yields in Rio Grande do Sul were more than 4 tons; in Goiás, by contrast, they still stagnated near the 1-ton level.[66] From a productivity standpoint, upland rice in the cerrado falls far below the mark. In 1988, Rio Grande do Sul produced a third of Brazil's rice on just 13 percent of the country's rice acreage. Goiás, then

Brazil's second largest rice producer, accounted for almost 20 percent of the crop's acreage but produced just 14 percent of the country's rice. Since then, the state's rice production has declined, mainly because Goiás lost the Tocantins region, which became a separate state. For 1992–93, the combined Goiás and Tocantins total averaged 820,000 metric tons. That is still second place, although far behind Rio Grande do Sul's 4.8 million metric tons. Average yields were much lower too, 1.6 metric tons per hectare compared to 5.1 tons in Rio Grande do Sul.[67] What accounts for the yield gap, and why do Brazil's farmers keep planting upland rice?

The National Rice Center

The strategy for upland rice depends on how favorable conditions are for production. According to Reinaldo de Paula Ferreira, who studies the cerrado's rice zones, the definition depends primarily on rainfall.[68] "A favored zone," he noted, "has dependable rainfall throughout the growing season. Even upland rice needs lots of water." The closer a region is to the Amazon Basin, as in northern Mato Grosso and in the new state of Tocantins, the better conditions tend to be. Farmers with a backup sprinkling system—circular pivots are the most popular—also have a good environment for upland rice, even in zones subject to periodic droughts. In 1990, however, the cerrado had only about 100,000 hectares with pivot systems.

The main aspect of "unfavorability" is a dry spell during the rainy season that lasts a week or more. In much of the cerrado, this happens most years, particularly in Goiás. "We can expect a dry spell of up to fifteen days in January," Luís Fernando Stone, who worked on water-use technology at CNPAF, said. "The result is a drop in yields commensurate with how long the drought lasts."

The cerrado also produces irrigated rice, much of it along the Formosa River in the north of Tocantins state. Sônia Milagres Teixeira, a rural economist at the rice center, told me about a cooperative of thirteen farm families that purchased 5,000 hectares in 1978 and started irrigated rice production.[69] A decade later, the Formosa Region had three cooperatives and a total of 60,000 hectares in irrigated rice. Farmers used modern varieties, applied fertilizer, and obtained yields that averaged 4.5 tons per hectare, on par with Santa Catarina and Rio Grande do Sul. After rice, they rested their fields or rotated to soybeans.

Given the tropical climate and constant day length, modern semidwarfs do very well in the cerrado when irrigated. For upland rice, the objective is to increase productivity in favored areas and stabilize yields where conditions were unfavorable.[70] Rice farmers without irrigation—the vast majority—seldom grow modern, semidwarf varieties, even where rainfall is dependable. Stone explained why: "Modern varieties are bred for irrigation, when not in standing water, they are very susceptible to blast fungus or get overtaken by weeds."[71] Consequently, most rice farmers planted upland varieties. With plenty of rainfall and some fertilizer, yields can reach three tons a hectare. Farmers, however, have to be careful with fertilizer. Upland varieties are already 1.5 meters tall. Adding too much nitrogen produces a taller plant that topples easily in blustery rain storms. "What we need for favorable rainfed zones," Stone said, "is an intermediate type that grows faster than the weeds but does not get too tall, a variety that can do well in acid soils, that holds up under the great downpours of the rainy season, but which does not need saturated soil continuously."

For most of the cerrado, upland rice is a second-rate alternative, planted in marginal areas by farmers who invest as little as possible. "Here rice is a frontier crop," Stone said, "the first crop farmers plant after clearing the land. They use rice in a system geared to creating pastures. Since they expect to make a living from ranching and not from rice, they are less concerned about low productivity. Given the dry spells, yields are precarious anyway; it simply does not pay to invest in fertilizer or pesticides." The challenge now is to change the frontier mentality, to get farmers to pay attention to productivity. In favored zones, the key is improved varieties. Even in unfavored areas, however, there are ways to stabilize yields.

Deep Plowing

"We try to get farmers to place more value on rice production," Stone said, "to make better use of the land they have rather than clear new land." An example is deep plowing.[72] Cerrado soils sometimes suffer from aluminum toxicity, which causes rice to develop a wide but shallow root system. To help the plant's root system grow denser and penetrate deeply into the soil, CNPAF advocates deep plowing to a depth of 30 centimeters. Deep plowing controls weeds, distributes the soil's dry matter better, and helps retain moisture. When farmers do not deep plow, the rice plant's shallow root system makes it highly sensitive to dry spells.

Besides deep plowing, the rice center promotes rotations in which rice is just one component. The positive impact rotations have can be applied across rice environments, including irrigated rice. Similarly, research on diseases and insect pests benefits all the region's rice farmers.

Blast Disease

Blast is the cerrado's most persistent, widespread, and lethal rice disease. Spread by spores, the fungus is a problem almost everywhere rice grows.[73] The conditions conducive to its spread vary greatly. With its humid, cloudy conditions, the cerrado is a "hot spot" where blast races proliferate rapidly. To control fungus damage, CNPAF screens hundreds of rice varieties for blast tolerance. Many lines came from CIAT, which also does blast testing. "The difficulty with blast," noted Dr. A. S. Prahbu, a plant pathologist, "is that strains are so localized. Rio Grande do Sul has rices resistant to its own local blast types. When transferred to Goiás, however, such varieties turn out to be susceptible. Farmers can apply fungicides, but that is expensive and impractical."[74] The main line of defense is varietal resistance, and the rice center has promising lines to work with and production strategies to reduce blast damage. Born and educated in India, Prahbu came to Brazil in the 1960s as a specialist for the cerrado, a region whose winter dry season and summer rains have much in common with southern India's semiarid tropics. He found Brazil "a tolerant place of diverse cultures." So he stayed. At CNPAF he specialized in blast control.

To minimize blast damage, farmers are encouraged to plant early and not too densely, use seed treated with a fungicide, and plant early maturing varieties. Deep plowing also helps, as it keeps plants healthier during dry spells, making them less susceptible to blast. Farmers are urged to avoid excessive nitrogen fertilizer, which tends to increase blast sensitivity, and to apply it at least seventy days after sowing. Harvesting cannot be delayed, as blast disease makes the grain heads shatter prematurely—a delay means losses when reaping. "Applying fungicides," noted Dr. Prahbu, "is not worth it. Farmers have to spray just when the grain is ready to sprout, otherwise, the yields will not cover the costs. Farmers who do not apply fungicide at the right time, or correctly, will lose money." The best strategy is to plant rices with at least some blast tolerance and adopt precautionary measures.

How much the weather favors blast buildup changes each year. "Over the course of the growing season," Prahbu said, "we can count up the favorable and unfavorable days. Some years, blast adds up to a tremendous problem, and other years it does not." Dew and dampness encourage the growth and spread of the fungus. When temperature extremes between day and night are great, dew formation is at its height. The leaves become damp and covered with droplets. A hard rain, followed by a strong sun that dries things up, is not conducive to blast. On the other hand, a slow steady rain that lasts all day is very favorable.

Upland rice suffers the most from blast, a situation short droughts intensify because they weaken the plant. Pivot sprinkling helps control blast in upland rice. Farmers get the crop off to a head start, even before the rains come, and additional watering can be done as needed. For irrigation, the cerrado has many streams and rivers: water can be collected from small dams that fill up during the rainy season, or it can be pumped from underground. A pivot system for 75 hectares costs about $170,000. So when farmers invest in a sprinkling system, it is not just for the rice crop. During the dry season, they rotate to lucrative vegetable crops that need less water. They can harvest during the winter, the off season for such crops in populous southern states such as São Paulo. In southern Goiás, farmers install pivots mostly for their bean crop. When watered, bean yields increase to 2 metric tons a hectare, three times the national average. And beans are more lucrative than rice. So although the rice crop benefited, it was mostly a side effect of the rotation system.

Despite the potential for small-scale irrigation, the cerrado's water resources are largely unexploited. "Feasibility studies should be intensified," noted Prahbu, "particularly for the cerrado's micro basins. India irrigates some 55 million hectares. Brazil, a much wealthier country, irrigates only about 5 million. And Brazil has an immense, semiarid zone that irrigation could make many times more productive."

Insects

Upland rice in Brazil's cerrado is beset by a host of insect pests. Spittlebugs, termites, and stem borers are the top three, but the rank order of the damage each does changes from year to year. Upland rice is not sufficiently profitable to justify using pesticides. Protecting the rice crop depends on resistant varieties and better production strategies.

According to entomologist Evane Ferreira, an intense, hardworking woman in her early thirties, the spittlebug sucks at the bottom of the rice plant, turning the stems yellow and drying them out.[75] When the infection rate gets out of control, losses can be enormous. In flooded fields, however, the insect is not a problem—it cannot get to the rice plant's submerged stems. Termites eat away at the roots, so weakening the plant that a short dry spell can be lethal. Deep plowing helps limit the damage, creating a deep, dense root system that tolerates more insect damage than a plant with fragile, shallow roots. Its effectiveness, of course, depends on how severe the attack is. Stem borers take their toll too. They lay their eggs on the rice plant, and when the larva hatch, they eat the plant's leaves and bore through the stems. Because a heavy rainfall can wash off most of the eggs, how long a dry spell lasts has much to do with how severe stem-borer damage is.

Breeders at CNPAF select for traits that minimize the damage an insect can do. They look for vigorous varieties with a deep root system. The rice center also promotes treated seed, which reduces disease and pest attacks during germination. A plant that is healthy from the start has much more resistance. Treated seed is cheap, safe, and does not harm pest predators. Crop density has to be considered too. Excessive competition weakens plants, and crowding can intensify insect and disease problems. Farmers also have to consider what they plant near their rice fields. "Soybeans," Ferreira noted, "actually inhibit the proliferation of a major termite species. Maize and sugarcane, both of which are grasses, attract the same pests that damage rice."

Farmers can avoid some insects by planting earlier than is customary, allowing the rice to mature before many troublesome insect populations reach their height. In the cerrado, most rice is sown between mid-October, when the rains begin, and mid-December. October plantings, for example, escape spittlebugs; December plantings minimize termite damage. Even better, if farmers wait until January, they can escape both pests. "The difficulty," Ferreira pointed out, "is that January usually brings a dry spell, which can wipe out young seedlings. To plant in January, farmers need a backup sprinkling system, a luxury for a crop of upland rice."

Coordination

Compared to irrigated rice in southern Brazil, upland rice in the cerrado presents a very different picture. The varieties used differ, as do production practices, inputs, diseases, yields, and profits. In Rio Grande do Sul, the state's rice agency worked with the crop because of its prime importance to thousands of farmers. In the cerrado, research at CNPAF focused on upland rice and rotations: either to pastures or to different crops. To make rice more profitable, and hence more important in its own right, upland yields have to increase. This is possible in favored areas, but seems less likely in unfavored ones.

Dr. Emílio da Maia Castro was director of Brazil's National Rice Center.[76] A serious, direct man, he knew the center's strengths and weaknesses. "We put so much effort into upland rice," Castro said, "because it withstands the deluge that comes at the height of the rainy season. So upland rice is about the best option. Of course, irrigated rice is far superior from a yield standpoint. And where conditions are favorable, we promote it." According to Castro, CNPAF acting alone cannot have a national impact. Brazil's topography and climate are simply too diverse. Unless states also have research programs of their own, there will not be a market for national research. "Our customers are as much state agencies," he said, "as local farmers."

Brazil's rice center (vis-à-vis state programs) and CIAT (vis-à-vis national programs) have structurally similar problems: how to make their research relevant to local rice production and how to transfer the results. For at each level in the research-extension scheme, whether international, within a country, or locally, an analogous set of practical problems can be found.

Research on blast disease illustrates the pitfalls of centralized research. The conditions that favor blast, and the precise strains involved, change from place to place. For its blast research to have an impact, CNPAF needs help from state programs, which knew more about the local situation. Working with state programs requires diplomatic skill. Local research scientists are often as well trained as those at CNPAF; they do not like meddling by outsiders and often consider "coordination" a polite word for control. How to balance the research advantages a national center has with the expertise of state programs and local extension is perplexing. For there is no obvious, general solution, only specific ones based on the research problem at hand.

Priorities for genetic improvement can be complementary and reinforcing, even when not identical. A state program bases its breeding objectives on local soil, disease, pest, and production problems. It cannot subordinate its interests to a national center any more than Brazil's CNPAF should do research for CIAT with little relevance nationally. Emílio de Maia Castro favored independent, parallel lines of research: "If some do not work out there is a chance that others will. This sometimes creates duplication, but if we overcentralize, we run the risk of following an incorrect line of research costly to everyone."

On-site testing exemplifies mutually beneficial, parallel research. Because Ceará State has only a small rice program and only one site for testing, the validity of its results is frequently in question. But when other states test the same variety and the results confirm the trials done in Ceará, it reinforces the conclusions drawn locally. The Ceará rice program can then recommend the variety to farmers with greater confidence. "Without coordination," Castro complained, "everything happens in a jumble. Different states release the same variety but use different names, which leads to tremendous confusion."

How Brazil's CNPAF works with state programs depends on their sophistication. It does not waste time on new varieties for Rio Grande do Sul; the state has a different climate and production system, and its rice agency has an excellent breeding program. Many state agencies, however, lack such research capacity and depend on CNPAF to do the time-consuming initial crosses and screening. State programs then select promising lines for use in their own genetic improvement programs. States send personnel to the national rice center for general training or to master a new technology such as tissue culture. The national center also sponsors workshops on problems common to several states, such as iron toxicity in the soil or red rice intrusion.

A state program is under pressure from growers for immediate results. Brazil's national center, by contrast, can have a long-term agenda. Biological control of plant diseases and integrated pest management are examples. Farmers want the kind of quick-fix that fungicides and pesticides promise. The national center, by contrast, can take the long road to environmentally safe technology. And it can afford to screen thousands of new varieties for blast resistance, passing on the best lines to state breeding programs.

Rice production in Brazil can be an agroindustry or a small-farm, family operation. Given the diversity of production systems and rice environments, coordination has to be carefully organized. Otherwise, the results will not get to the right place, or the right people, on time. Consequently, CNPAF takes a regional approach to problems such as genetic improvement. This promotes a critical, cooperative exchange based on actual field experience and saves on research funds.

Crisis

Brazil's research system is supposed to link a national center to state programs and local extension. For example, CNPAF has a budget to fund research projects managed and executed by state programs. Its division for technology transfer works with state-level extension to set up demonstration plots, organize training sessions, and present results to producers, especially cooperatives. Such is the ideal, at least on paper. In practice, Brazil's financial instability took its toll. For example, twenty full-time researchers in rice, many of them with doctoral degrees from U.S. universities, worked at CNPAF. But by 1990, triple-digit inflation and political disarray had undercut what critics considered Latin America's best agricultural research program. Budgets shrank, the purchasing power of salaries dropped by half, and the national extension agency was scrapped. Center directors such as Castro struggled to keep their budgets and research teams intact.[77]

The situation was worse for state programs. Waldemar Pinto Cerqeira, director of the Agrarian Research Agency in Goiás (EMGOPA), explained the predicament many states faced.[78] "Our research budget is so tight," he said, "we lease out personnel to projects funded by private firms, producer groups, and multinationals. They have better equipment, more land, and more backup labor. Without such private support, we could not survive." For example, the agency researched pineapple production for canning, a justifiable project, as the pineapple industry brings investment and jobs to the state. But the private sector rarely invests in long-term, valuable projects such as biological pest control. Waldemar believes that state programs need more autonomy. "The national centers," he said, "can finance projects and coordinate research but they cannot have a local impact without strong backup from state programs."

SOLUTIONS

Research at the national and state level are supposed to be mutually reinforcing; the results, promoted through extension, are expected to benefit farmers. Such principles are easy to write down but difficult to execute. Designing research projects so they end up helping farmers is a great challenge. How this happens with rice in Brazil, as opposed to work in Asia and at IRRI, shows both the pitfalls and the prospects involved.

Rice production in Brazil poses problems that are marginal or irrelevant at IRRI. In Rio Grande do Sul, red rice intrusion is so serious that it is central to the state's approach to rice: from breeding objectives to seed production and rotations. That said, some parts of the state suffer less from red rice than others. A prime factor is the scale of production. In neighboring Santa Catarina, where small farms predominate, red rice is never mentioned. And in Rio Grande do Sul, it is in zones of large-scale production that red rice is most troublesome. To eliminate it means rotations: to newly planted pastures, to soybeans, or to wheat, all of which are large-scale solutions. In short, Rio Grande do Sul's rice problems have little in common with how IRRI, given its experience in Asia, thinks about the crop.

Asian rice production is dominated by small producers whose tiny plots defy Brazilian ideas about economies of scale. In Asia, multiple cropping, azolla, green manure, and mung beans are all part of the rice vocabulary. In Brazil, the language of rice is so different it is sometimes hard to remember it is the same crop. At CNPAF, upland production dominates research. At IRRI, upland rice is secondary, pursued in the name of environmental protection and germ-plasm collection. In the cerrado, rice alone rarely provides farmers with a livelihood; they depend as much on ranching or on winter vegetables and beans. For Asian rice farmers, rice is much more important than other crops; everything else is secondary. Even the list of pests and diseases is different. In the cerrado, spittlebugs and blast rank much higher than the plant hoppers or tungro virus of Asia.

Given such contrasts, it is not surprising Latin America has a regional rice-improvement program, an effort that CIAT coordinates. And it is not surprising Brazil has a national center with one set of priorities and state programs with another set. Even within a state, rice production can warrant further subdivision. Rio Grande do Sul has four distinct rice-growing regions, and within each there are county offices to organize extension and monitor seed production.

The lesson is that despite great differences, common action is possible. What matters is respect for the integrity of each link in the chain of action. Such respect is hard to build and hard to maintain. In Asia, where vast populations depend on rice, production is an affair of state. The links between rice research and on-farm application get much more attention than in Latin America. Brazil has an excellent research system, but investment is concentrated at the top, in federally supported, national research centers. To gain the most from that investment requires strong research allies in state programs, whose impact, in turn, depends on extension. In agriculture, the puzzle cannot be solved with just one piece. For success, all the pieces must be in place.

CHILE

3

I crossed over the Andes into Chile by bus from Argentina and headed south toward Puerto Montt. Because of the heavy Pacific rains that drench the Chilean side of the mountains, southern Chile is temperate, damp, and lush. Some 3,000 miles long from north to south, yet never more than 100 miles wide, Chile is unique geographically. Great differences in latitude and elevation combine to make it a diverse place agriculturally.

Regions in Chile are so different in climate as to form almost separate countries. The North begins at Arica near the Peruvian border and extends through the Atacama Desert to La Serena in Coquimbo Province—a distance of almost 1,000 miles. Except where shallow rivers break through the Andes to the Pacific, it is too hot and dry for agriculture. Only about 10 percent of Chile's fourteen million people live there.[1] Most of Chile's population, almost 75 percent, is concentrated in the country's great central valley. It extends from just north of Santiago, the nation's capital, to Temuco, 400 miles south. With more than five million inhabitants, metropolitan Santiago has 40 percent of the country's population. Beyond Temuco begins Chile's South, which includes the lake district and the provincial capital of Puerto Montt. Puerto Montt is connected to Santiago by the central highway and the railroad. To go farther south—it is still 1,000 miles to Punta

Arenas on the Magellan Straits—there is an ocean passage from Puerto Montt through the country's antarctic archipelago.

From the central valley south to Puerto Montt, Chile is a nation of temperate climates, but its seasons are the reverse of North America's. Chile sends its summer raspberries and table grapes north to U.S. markets in January and February. I arrived in Puerto Montt on 10 March 1990. That far south, fall was in the air; the high for the day was 62 degrees Fahrenheit. Farther north in Santiago, it was still summer and 83 degrees.

Southern Chile is cool even in summer, an ideal place for dairy and potato farming. It is rainy; Puerto Montt gets more than 2,000 millimeters a year. It rains every month, with some 1,200 millimeters falling during the autumn and winter: from April through September. The Santiago region, by contrast, is much drier, with just 350 millimeters of rain, almost all of which falls between May and August.[2] Wheat production, vineyards, and fruit trees do well in this drier part of Chile's central valley; irrigation from lakes and mountain streams supplement the rainfall. In the far north, it never rains.

WORKING WITH FARMERS

That farmers should be included in research projects is a premise invoked almost everywhere. Much of the time, it means farmers do what researchers tell them. At IRRI projects, for CIAT, and in Brazil, farmers tested new varieties, tried out different production methods, and worked on pest control, usually for researchers. On-farm trials are undeniably important. All too often, however, farmers are included only at the end of the project. What matters most to professionals is the quality of the research, a judgment made by colleagues, not by farmers. Even with high-yielding modern varieties there are "no easy harvests." As experts have argued for years, a successful production program requires strong local organization and on-farm education; research alone is insufficient.[3]

I did not go to Chile exclusively for rice. As dietary staples, wheat and potatoes rank far ahead. Chile is self-sufficient in rice, but total production for 1992–94 was just 133,000 metric tons; per capita consumption of milled rice is only 10 kilos. A minor crop, rice gets only a small share of the country's agricultural research budget. My interest in Chile was that in research it treated scientists and farmers as colleagues. Chile's National Agrarian Research Institute (INIA) required its scientists to be part of farmer-oriented, local research groups. I wanted to know how these groups worked and how representative they were.[4]

"Given our topography," said Luis Becerra, "it is difficult to apply solutions beyond a specific microclimate."[5] Becerra helped organize on-farm research groups for the Quilamapu Research Station near Chillán, capital of Nuble Province. The Chillán region has a warmer climate than Puerto Montt, 380 miles to the south, and it does much better on rainfall than Santiago, 250 miles to the north. The district served by the Quilamapu Station is a quilt of microclimates staggered from east to west: the Pacific coast, the central valley, the foothills, and the snow-capped Andes. Rainfall averages 900 millimeters along the coast, but tapers off to 300 millimeters in the foothills. The district's agriculture includes cereals, beans, fruits, ranching, and dairy farming. And new crops, especially raspberries, kiwi, and sugar beets, are making headway. The Quilamapu Station lacks the facilities, staff, and budget to work intensively on so many crops and agricultural systems. "Without the collaboration of local farmers," Becerra said, "a small research station can accomplish little."

Chile involved farmers in research only after the country's agricultural output had collapsed. Between 1968 and 1978, per capita food production dropped by 15 percent. As a result, Chile's food import bill increased enormously, eroding its trade balance. Between 1980 and 1984, food and beverage imports accounted for more than 13 percent of Chile's total imports—compared to just 4.5 percent in neighboring Argentina.[6] To stop the decline, Chile's Agrarian Research Institute was reorganized in 1982; this time, on-farm research and technology transfer were basic components. The objective was self-sufficiency in rice, wheat, sugar, oils, milk, and meat production.

Compared to 1977–79, total agricultural production in Chile for 1987–89 had increased by a third; per capita output was up almost 12 percent. The food category had dropped to just 4.3 percent of Chile's total imports in 1986, and for 1987–88, the country had a favorable trade balance in agricultural products of more than $1.2 billion. Chile exports table grapes, raspberries, and other fresh fruits to the United States and Europe, taking advantage of the winter market in the Northern Hemisphere. It is a remarkable achievement.[7]

"The recovery of Chile's agriculture," Becerra said, "is in large part due to the Technology Transfer Groups." In 1990, more than 120 groups were linked to INIA's five research stations. Groups have about fifteen members, most of them influential local farmers. When a group is first organized, INIA provides a coordinator for technical assistance. Coordinators work with the groups on a part-time basis, helping set up on-farm trials to test new varieties and production methods. In the beginning, a group's success depends a lot on its coordinator, but as a group gains experience and confidence, it becomes more self-reliant. Eventually, groups cover the cost of technical assistance or hire their own agronomists.

"The best groups are independent and set their own priorities," Becerra said. "They meet regularly to critique the trials researchers are doing. Evaluations can be tough. Farmers want to know what works and what does not, and why. The members have to justify their own practices too, not simply criticize new technology." Some groups in Chillán have eight years or more of experience and are completely self-financed.

Groups are composed primarily of substantial farmers. Nonetheless, considering the crops they produce, the quality of land members own, and whether they have irrigation, groups are quite diverse. In the Chillán district, group membership crosscuts microclimates, crops, and production systems.

Carlos Lago, director of the Quilamapu Station, told me that everyone has to work with the groups, even the geneticists in charge of breeding programs.[8] They have to defend their research at meetings with farmers, justifying their breeding strategies. "Such give and take," said Lago, "has helped scientific investigation enormously. Groups help us get new varieties into production, and promote better crop management. They have legitimized our research by putting it into practice." Such a connection is very important politically. INIA is now in direct contact with its customers, with people who understand how research benefits them. "Compare this to Chile's universities," Lago said. "They do agrarian research too, and they have contributed a lot to Chile. But they cannot show it; they have no organized groups in the field with a stake in the university's research budget."

Making on-farm critiques part of research had much opposition initially, especially from research specialists. "I did not think technology transfer was worth it; I considered it a waste of time," Mario Mellado told me.[9] Mellado headed Chile's wheat program and worked at the Quilamapu Station. "I felt my job was breeding exclusively," he said, "to genetically adapt new varieties to our soils and microclimates. I was wrong. Working with the groups is the best way to get new varieties into the field. Without their help, Chile's wheat program would not be a success."

Wheat

Chile is a great wheat-eating country; for 1989–91, its per capita consumption of bread and wheat products averaged 135 kilos, considerably more than the U.S. average of 115 kilos and almost three times more than Brazil's 44 kilos. During the 1970s, however, production and yields dropped and wheat imports climbed. In the period 1964–68, Chile's wheat production averaged 1.2 million metric tons a year, with yields of 1.6 tons. For 1979–83, average production dropped to 775,000 metric tons, with yields of 1.7 tons, hardly any yield improvement. The worst year was 1983. Chile planted only 360,000 hectares of wheat for a total harvest of just 590,000 metric tons. Yields were only 1.6 tons per hectare. Between 1981 and 1983, Chile imported more than 3 million metric tons of wheat at a total cost of almost $670 million.[10]

Given the variations in rainfall, temperature, and elevation between Santiago and Puerto Montt, Chile can plant wheat from April into September. But it needs an assortment of winter and spring wheats. Working with agronomists in different zones, breeders improved local varieties by crossing them with modern ones, creating what the program called "alternative" wheats. Or they introduced modern varieties from countries with climates similar to regions of Chile. "Farmers needed many options," Mellado noted, "given our microclimates, we could not rely on a couple varieties to do well everywhere."

"In 1989, Chilean farmers planted twenty-one different spring wheats, sixteen alternative wheats, and three winter wheats—all of them modern varieties," Mellado told me. By then yields had almost doubled. For 1992–94, Chile's wheat production averaged 1.4 million metric tons on 406,000 hectares for an average yield of 3.4 metric tons per hectare.[11] In fact, Chile is almost self-sufficient in wheat. Mellado said Chile could easily produce the 1.8 million tons of wheat the domestic market needs but lacked sufficient storage capacity.

Rice

The Quilamapu Research Station stressed both genetic improvement of crops and crop management. The station is in the midst of Chile's fertile central valley, the heart of the country's most traditional farming area. Between Talca and Temuco, a greater percentage of the population lives in rural areas than in any other part of Chile. In 1990, almost 85 percent of the region's producers were small farmers. Although their holdings made up only 30 percent of the region's total, it included half the irrigated farmland.

Most of Chile's rice, about 70 percent, is grown just north of Chillán in Linares and Talca Provinces. According to Roberto Alvarado, breeder for

Quilamapu's rice program, most rice farms have about 10 hectares.[12] He told me that farmers plant about 40,000 hectares of rice a year. "That's not much, but in good years it is sufficient for the domestic market, for self-sufficiency." Chile irrigates almost all of its rice. Soils have sufficient clay content to enhance water retention. Farmers directly sow pregerminated seed into a soaked field; they do not transplant their rice.

Rice in Chile does not warrant a full-fledged research program with the standard specialties. Consequently, Alvarado's breeding efforts have to be backed up by a multidisciplinary team that studies disease and production problems for several different crops. The rice zone's temperate climate restricts the selection of rice varieties to those with cold tolerance. In the 1960s, the most successful rice, called *oro* or gold, came from Italy. It is shorter than local rices and more responsive to fertilizer. During 1965–68, yields averaged 2.5 tons per hectare. Prior to this, yields were too low for profitable commercial farming.

The advantage of Chillán's summer climate is that low humidity reduces the risk of fungal and virus diseases. In the fall, when the rainy season starts, cool temperatures inhibit most diseases and minimize pest problems. The disadvantage is that early frosts pose a threat to rice production. According to Roberto, early maturing semidwarfs solved the problem: "Because they can be harvested in 130 instead of 180 days, farmers can delay planting until later in the spring, from mid October into early November, and harvest earlier in March at the end of summer." The new rices are indica types that are long-grained, fine, and hard: the cooking characteristics Chileans prefer. For 1985–87, yields averaged 3.9 metric tons per hectare; for 1992–94, 4.4 tons—a big improvement.[13]

To benefit fully from a new variety's potential, farmers need to use the production system most suitable for it. "Farmers think a new variety will solve all their problem," Alvarado said, "but the key to success is as much in how someone farms as it is in the rice plant's genetic makeup." In the Chillán region, poorly constructed dikes make it difficult to control the water level in rice paddies with sufficient precision. The result is weed intrusion into shallow fields. "Yields on many small farms are stationary," said Alvarado. "To increase further, they will have to improve soil preparation and water control." Soils are not very fertile. So farmers add nitrogen as well as small amounts of phosphorous and potassium. As for water control, the land has to be graded better at planting time and dikes have to be roughly the same height. When one field has 300 millimeters of water and another only 50 millimeters, yields are erratic.

Chile's market for rice depends on how much people eat. While per capita consumption has increased from 5 to 10 kilos since 1960, the pace is slow. Compared to Colombia, where rice consumption is concentrated disproportionately among the poor, in Chile it is the middle class that has taken to rice. Consumption might yet go up a couple kilos, but the rise is unlikely to be sudden or dramatic. So although Chile has more land suited to rice than it uses, market prospects do not favor additional production. In fact, as a hedge against a rice glut, and to rest the soil, the Quilamapu Station encouraged crop diversification.

"Farmers with lots of land can rest their rice fields between crops," Roberto said. "They leave it fallow to revegetate naturally. Small farmers, however, have no land to spare. So they plant rice after rice, which wears out the soil." The solution is to rotate from rice to other crops, but such substitutes have to be profitable. The alternatives under trial in 1990 included sugar beets, sunflowers for cooking oil, watermelons, and cereals, notably wheat and maize.

How did the Quilamapu Station get farmers to adopt new rice technology and rotate to different crops? Part of the answer is in the work it did with local organizations, especially the on-farm groups.

The Parral Group

Roberto Alvarado arranged for me to attend a local meeting at Parral, a small town about 20 miles north of Chillán.[14] I went with Antonio Valdez, an agronomist from the Quilamapu Station who helped the group set up its on-farm trials. We met in a small café. The group's leader, José Luis, outlined its activities.

The Parral group, José Luis told me, was organized five years ago. It has strong ties with researchers at the Quilamapu Station, who work jointly with them on demonstration plots. In rice, the group promotes better weed control and improved crop management. The group is now self-supporting; in fact, it pays Antonio Valdez for the technical direction he provides. The group's fifteen members have an average of 40 hectares in rice; a couple of members have a total in excess of 500 hectares. This makes the group atypical, because most rice farmers have much less disposable land overall and only about 10 hectares in rice. "That is why we all have to stick together," José Luis told me. "As individual producers we have little bargaining power. Small farmers in particular end up selling at harvest time when prices are the lowest. That is why joint marketing and storage is so important."

The Parral group exchanges ideas freely and critically. Over the years, members have become good friends; they trust one another. The group meets at least once a month at a member's farm on a rotating basis. It assesses how well a member's crops are doing and evaluate any trial plots. The group identifies mistakes and poor practices. On occasion, invited speakers talk about rice problems or new production technology. When the Quilamapu Station organizes a trial on a member's land, researchers have to justify the test to the group.

Given low prices and serious weed problems, the group wants alternatives to rice. "We need to know which crops can do well in soil with such a high clay content," José Luis emphasized. "We can plant rice on the same land for a couple of years, but then we have to rest the soil, ideally, for three years. That is a long time. What we need are alternatives to taking land out of production." Thus far, the group has experimented mostly with beets, wheat, and beans. A breakdown of production by crop for the region around Parral shows that beets have gained as an alternative to wheat. For the 1987–88 season, the region planted 960 hectares in beets and 2,230 in wheat—compared to 10,780 hectares in rice. For 1988–89, however, beets went up to 2,490 hectares whereas wheat fell to just 850; the rice acreage held relatively steady, increasing slightly to about 11,000 hectares. What matters is that the region's farmers, led by the Parral group, tried alternatives.

Coffee finished, we headed to a member's farm where the Quilamapu Station had demonstration plots on fertilizer use. The soil specialist from the station came too; he wanted to see the group's reaction. It was March, the end of summer in Chillán and harvest time for rice. Many fields were drained; it was hot and dusty. We took gravel roads as far as we could and then continued by foot across marshy fields and irrigation dikes.

The first trial plot was so heavily fertilized that the weak-stemmed, top-heavy rice lodged. Plots with less nitrogen plus a little phosphorous did much better. The lesson is that a balanced fertilizer is best for a healthy, resistant plant, and that farmers can cut back on the quantity used. How to apply fertilizer to greatest effect was likewise discussed. Experienced showed it is best to incorporate it into a muddy field prior to seeding, a conclusion long since reached in Asia.

Small Farmers

Farmers of substance and standing in their communities joined groups like that at Parral. But what about Chile's small farmers? Carlos Altmann directed the technology transfer division at INIA's headquarters in Santiago. A soft-spoken man in his late fifties, Altmann was direct and candid. He told me that small producers accounted for almost a third of Chile's agricultural output.[15] They were important not only in the rice sector but also in cereals such as wheat and maize, and even in the export sector, particularly viticulture. Nonetheless, INIA's group approach had not been applied to small farmers. Instead, the available technical assistance came from a separate agency, the Agrarian Development Institute. With funds from the Inter-American Development Bank and the Chilean government, it was setting up twenty demonstration centers for technology transfer. Each center was to work with groups of small farmers; an agronomist was assigned to each group.

"For Chile's small producers," Altmann noted, "there is an agrarian solution. They can have a better life based on agriculture. For families without land who work as sharecroppers or day laborers, there is no agrarian solution." Chile had undertaken a major land distribution program in the 1960s, and even though some consolidation had occurred, the Pinochet regime did not undo what the reform accomplished. Consequently, as Altmann said, "Chile has many small producers with 25 to 30 hectares. For them, family incomes can be stabilized and improved. But there are many rural workers outside the reform, families with just a house and a small plot. For them there is no agrarian solution."

Chile needed to work more with its small farmers. And it already had a way to do this—the Technology Transfer Groups. The new regime promised to pay more attention to the country's small-farm sector; it intended to use the group approach with small farmers. In fact, INIA proposed this at the start—that the group structure should radiate out to farmers of all types.

"To work successfully with small farmers, we have to recognize that their situation is different," Altmann said. "Agrarian reform often presumes those involved have an entrepreneurial mentality. That is not always so. Small farmers do not like credit; they are afraid they might lose their land. Participation in a group can help change this mentality. It gives farmers the support they need to take risks, to make credit an acceptable tool of development."

In many ways, Chile's INIA already helps small farmers. It works with the Agrarian Development Institute to make sure small farmers get improved varieties and high-quality seed. It trains extension workers, organizes special courses, and sets up demonstration plots. Nonetheless, small farmers face considerable obstacles. "It costs them more to change," said Altmann, "and the pace of change is slower. A different production system can require a change in habits that are part of daily life."

"To work successfully," Altmann said, "it is important to ask just exactly where one wants to start. Small farmers only participate in a group for a clear, understandable reason. A group has to have a work plan. Farmers want to see how something works in practice—not just talk about it. That is what makes demonstration plots so important. New technology has to be affordable, it has to be easy to implement, and the benefits must be well established. What small farmers need most is no secret. They need training and credit, a dependable market, and stable prices. They need to see there is a way out."

RATIONALE

In monsoon Asia, rice production is so linked to the public welfare it is almost a matter of national security. In Thailand and Indonesia, rice accounts for more than half of the calories people consume daily; in Indochina it is almost 70 percent. Per capita rice consumption ranges from 128 kilos in Thailand to 190 kilos in Burma.[16] That most Asian governments endorse modern varieties, invest in rice research, and promote extension is not surprising. But what happens in countries in which rice is a secondary crop of limited importance?

Chile demonstrates that even a small rice-producing country can benefit from high-yielding modern varieties. To understand how this happened, rice has to be viewed in its context. The approach Chile took to rice is part of a larger strategy to shore up agricultural production. Like many countries, Chile could afford only a modest investment in research, with little specifically for rice. By all accounts, what made that investment pay off so well was the work done by the Technology Transfer Groups.

The role the groups play in research reduces the time and cost involved in developing new varieties. The outstanding example is wheat. Success depends on testing and introducing many new varieties simultaneously, varieties that must be adapted to the country's complex of wheat microclimates. The impact the groups had was mutually reinforcing; it was not confined to a specific crop. In the Parral Group, sustainable agriculture through crop diversification became an ongoing research focus largely because farmers wanted it that way.

In Chile, using research funds efficiently is not simply a matter of cost-accounting at district headquarters. The value of each peso invested was enhanced by the work of Technology Transfer Groups all over Chile. Nonetheless, as Carlos Altmann warns us, the groups are not a recipe for action under all circumstances. To be successful with small farmers, groups need to start with more modest, practical objectives.

4 COLOMBIA

Colombia is Latin America's second most populous Spanish-speaking country. It had 36 million inhabitants in 1994, far behind Mexico's 92 million but just ahead of Argentina's 34 million. It is also Latin America's second largest rice producer, after Brazil. Yields for irrigated rice are 5 metric tons per hectare, impressive even by Asian standards. Per capita consumption of milled rice was 32 kilos in 1991 and is expected to be 40 kilos in the year 2000.[1]

RICE IN COLOMBIA

Colombia introduced IR8 in 1968, making it the first Latin American country to do so.[2] In southern Brazil and in Chile, farmers needed a cold-tolerant rice. Because the first modern rices were indica types sensitive to cold weather, they could not be grown successfully without modification. In Colombia, the country's rice zones do not have cold spells, much less frost. So it was easier to switch to modern indica rices. Nonetheless, semidwarf lines from Asia were not ideal; IR8, for example, was more susceptible to local diseases and pests. Farmers needed high-yielding lines better suited to Colombia's soils, pest problems, and blast races. To work on this, a joint breeding project was set up involving CIAT and the Colombian Agrarian Institute (ICA).[3] In 1971, the ICA-CIAT program released CICA-4, which had more disease resistance and better grain quality than IR8. Three years later, more than 90 percent of Colombia's irrigated rice acreage was planted to improved, semidwarf varieties.[4] The new rices spread quickly because the Colombian Rice Growers Federation (FEDEARROZ) promoted them—along with the associated technology.[5]

In Asia, semidwarf varieties did best when fertilized and irrigated. So too in Colombia. Modern varieties first spread along the upper reaches of the Magdalena River, where farmers already irrigated their rice.

Impact

Situated between two mountain chains, the Magdalena Valley is one of Colombia's richest agriculture zones. The Magdalena River rises outside the mountain town of Pasto near the Ecuadorian border and flows north for 700 miles before reaching the Caribbean. To one side is the Cordillera Oriental; Bogotá, the capital, is atop this rugged chain. To the west, the Cordillera Central juts up, Medellín its principal city. In Tolima and Huila, departments in the upper Magdalena Valley, farmers use gravity-fed irrigation systems to channel water from the river's many tributaries to their rice paddies. During 1964–66, prior to the introduction of semidwarf varieties, farmers irrigated on average some 120,000 hectares of rice; yields were close to 3 tons. A decade later, for 1974–76, average yields exceeded 5 tons. By 1980, Colombia's irrigated rice acreage had increased threefold to 340,000 hectares. Total production rose in tandem, from an average of 650,000 metric tons for 1964–66 to almost 2 million tons for 1981–82. By then, rice prices had dropped and Colombian consumption had doubled. Rice became Colombia's most important food crop, with consumption disproportionately concentrated in low-income households.[6]

Farmers who could irrigate their rice made up for the price decline with higher yields. This was not true for farmers whose rice crop depended on rainfall. As in Brazil, modern varieties widened the yield gap between the two production systems. From 1964 to 1966, for example, almost two-thirds of Colombia's rice crop was produced under rain-fed conditions. Back then, the difference in yields was not so great. For irrigated rice, yields averaged about 3 metric tons per hectare; for rain-fed rice, 1.3 metric tons: a 2 to 1 advantage. But once the rice revolution set in, the advantage shifted so decisively that rainfed production dropped sharply. For 1974–75, the average yield for irrigated rice was 5.3 tons; for rain-fed rice it was only 1.5 tons: a gap of almost 4 tons per hectare. No wonder farmers converted to irrigation wherever possible. In 1981–82, Colombia irrigated more than 75 percent of its rice. By then, the acreage devoted to rain-fed rice had dropped by half, from 219,000 to 107,000 hectares.[7]

Meta

The Department of Meta is across the mountains to the east of Bogotá. A narrow strip of fertile piedmont soil stretches north and south along the foothills, which slope down to the *llanos*, Colombia's savanna grasslands. Villavicencio, Meta's capital, is at the edge of the piedmont just 70 miles from Bogotá. Land is cheap, the soil fertile, rain abundant, and a paved road, albeit narrow and dangerous, provides access to the Bogotá market. The region is well suited to agriculture, but in the 1960s it was still a frontier zone of cattle and cowboys.

Just as IRRI works on new varieties for rain-fed environments, so too does CIAT. In fact, Meta's piedmont belt gets so much rain, over 3,700 millimeters between May and October, that varieties suited to irrigation can also do well. Beyond the piedmont, soils become acidic, often with a high aluminum content. And rainfall, although heavy, tapers off from west to east. Santa Rosa, a test site 15 miles east of Villavicencio, still gets some 2,500 millimeters a year, but Puerto López, another 45 miles east, gets only 2,000. So as the rice zone moves farther into the savanna, conditions become drier and upland varieties do better than either irrigated or rain-fed rices.

The savanna has less rain than the piedmont, but at least it is dependable. In Brazil's cerrado, also an upland rice zone, a dry spell always looms as a threat to production. Short droughts rarely occur in Colombia's savanna. Consequently, the zone has unusually favorable conditions for upland rice. And the Colombian government has backed savanna rice improvement as part of its development policy. In fact, ICA has its main center for rice research at La Libertad, near Santa Rosa. Darío Leal, a large, enthusiastic man, headed the agency's rice program.[8]

A decade ago, the Magdalena Valley dominated the country's rice production, but since then, Meta has gained ground. According to Leal, in 1988, Meta planted 120,000 hectares of rice, which accounted for a third of Colombia's production—compared to 42 percent for the Magdalena Valley and 25 percent for the north coast. In fact, Villavicencio's urban development is based on rice. I saw the evidence everywhere I went in town: the equipment repair shops, supply stores, silos, rice mills, and storage depots with tons of nitrogen fertilizer. Producing seed for Meta's rice farmers is likewise big business. So too is shipping rice across the mountains to Colombia's big cities. The employment tied to rice adds up to thousands of local jobs. And to spur Meta's development, the government invests in hydroelectric power and road construction.

Rice research at the La Libertad Station includes both irrigated and upland rice. It has special programs for blast disease and weed control, serious problems everywhere in Colombia. "When you look at the data on yields per hectare," Leal noted, "Colombia gets only about 4 metric tons. But countries with high yields typically irrigate 90 percent or more of their rice. Colombia, by contrast, irrigates only 65 percent; it has a lot of rain-fed rice, much of it planted in upland varieties, which pulls down the average yield.[9] In Tolima and Huila, departments where almost all the rice is irrigated, yields are over 6 metric tons per hectare. So for irrigated rice, Colombia's yields are comparable to top producers like Korea and China."

Colombia has high yields, but also high costs. In Tolima, farmers use some 280 kilos of seed to plant 1 hectare of irrigated rice, twice as much as in Brazil. They use too much nitrogen fertilizer and spray pesticides on a predetermined schedule, regardless of the actual insect threat. Leal said that even big growers with high yields think a reduction in pesticides is too risky. "So we did a three-year study," Darío said, "in which farmers cut back on seed quantity, on fertilizer, and on pesticides. In almost all cases, yields held up despite the cutbacks. Costs went down and profits went up."

Working on genetic improvement is a task ICA shares with CIAT. Lines with higher yields or better blast tolerance come from CIAT, and ICA crosses such lines with local varieties. Subsequent regional tests are performed by the Rice Growers Federation, which also works on extension and monitors seed production.[10] Certifying a new variety for release, however, is the exclusive prerogative of ICA.

When I met with Darío Leal, Colombia was reorganizing its agricultural research, switching from regional to product-specific centers. Rice was to stay at La Libertad, along with soybeans and African palm. Nonetheless, how much the agency could do was limited by its modest budget. The Rice Growers Federation, by contrast, was a rich agency supported by a small tax on processed rice. Given two agencies whose functions overlapped, one with a precarious budget, the other with financial security, coordination was difficult, even under the best of circumstances. Consequently, research of vital interest to Colombia's rice growers gravitated to CIAT.

CIAT

Located in Colombia's fertile Cauca valley, rice is only one item on CIAT's research agenda. It also works on field beans and the rooty cassava plant, crops native to the Americas. Its research on tropical pastures seeks to reduce erosion and restore degraded grazing lands. Compared to beans and cassava, for which CIAT has worldwide responsibilities, its research mandate for rice is restricted to Latin America. So it is not surprising that CIAT's rice research reflects the region's problems. Red rice, rotations, and upland production are key issues for national programs in the Americas, and also for CIAT.

The initial objective at CIAT was to develop semidwarf varieties to replace Latin America's low-yielding traditional rices. Both CIAT and ICA were soon crossing IRRI rices with local varieties. By the early 1970s they had lines suitable for the region's tropical climates. A decade later, in Brazilian states such as Santa Catarina and Rio Grande do Sul, zones that require cold tolerance, breeders crossed CIAT-IRRI lines with local cultivars. Chile too used germ plasm from the two centers in its breeding program.[11] By the mid-1980s, CIAT had worked out a regional approach to germ-plasm testing with IRRI. An INGER subdivision for Latin America and the Caribbean was organized for rice evaluation. In the past decade, INGER–Latin America has contributed to the release of some forty-five new rice varieties.[12]

For Latin America overall, average rice production increased from 11 million metric tons in 1969–71 to 17.5 million tons for 1989–91, and to 19.1 million tons for 1992–94. Between 1971 and 1991, inflation-adjusted rice prices fell by 24 percent. In 1990, irrigated paddies accounted for 11.5 million metric tons of Latin America's rice, or about two-thirds of the total.[13]

Of the region's twenty-five rice-producing countries in 1990, only half had active rice-crossing programs.[14] So how advanced rice research is in specific instances differs greatly. Brazil's rice program has more experience than CIAT in key areas such as rotations and red rice. Oil-rich Venezuela, by contrast, neglected agriculture. It habitually imported food to make up for shortfalls in domestic production. Its rice yields per hectare dropped from an average of 3.1 metric tons during 1975–77 to 2.6 tons during 1985–87. Assisted by CIAT, Venezuela subsequently targeted its rice sector. Yields for 1988–90 improved to 3.3 tons; for 1992–94, they averaged 4.2 metric tons. Ecuador added to its rice area, introducing gravity-fed irrigation projects near the coast along the Guayas River. Between 1982–84 and 1992–94, its average rice acreage almost tripled, rising from 122,000 hectares to more than 340,000 hectares. Yields for 1992–94 were 3.5 metric tons.[15]

Thus, CIAT has a diverse clientele. In countries where rice research has fallen behind, it collaborates on breeding and production projects. Where research is advanced, it supplies germ plasm. Where countries share problems, as with red rice, it promotes networks. As national programs change, CIAT's approach to rice has to keep in step.

According to agricultural economist Luis Roberto Sanint, CIAT's rice program once concentrated almost exclusively on breeding for irrigated environments.[16] The rice program was a germ-plasm program; it did not even have an entomologist. "We now focus more on crop management," he said, "from land preparation and planting to harvesting and storage." The new emphasis includes weed control, new rotation systems, and reducing production costs.

Bob Zeigler, an American plant pathologist with prior experience in Africa, was head of CIAT's rice program.[17] A stocky, outspoken man in his early forties, Zeigler considers it of utmost importance how farmers manage a rice crop. He pointed out that weed problems often reflect water control problems, which can be traced to how the land is prepared, how the dikes are maintained, and to the machinery used—all of which reflect the availability of credit and all of which are production problems. To make up for poor production practices, farmers use more herbicides. "We could end up with herbicide resistance in our weeds," he said, "much as insects mutate to survive insecticides and end up stronger than ever." Rotations, by contrast, protect the soil, reduce weed control problems, and lesson reliance on herbicides.

To cut back on costs likewise meant changes in production technology, from fertilizer use to cropping systems and pest control. According to Luis Sanint, better crop management has helped farmers reduce their reliance on chemicals. "In 1980," he said, "Colombian rice farmers sprayed against insects and disease nine times per crop; in 1990, they sprayed three times or less."[18]

Upland Rice

Compared to yields of 5 metric tons per hectare for irrigated rice, upland rice seems second best. Because under the right conditions yields are good and the crop is profitable, CIAT continues to work on upland rice.

In Brazil's cerrado, upland yields rarely surpass 1.5 tons per hectare. The culprits, however, are low-fertility soils and a rainy season interspersed with droughts. For the cerrado, what constitutes a "favorable" environment depends almost exclusively on the rainfall pattern. In Meta's savanna, rainfall is adequate and well distributed; what constitutes a favorable environment depends on the soil's fertility. In the llanos to the east, soils are less fertile and more acidic. Farmers need acid-tolerant upland rices that do well without much nitrogen fertilizer. Considering the low costs of production—that farmers do not need irrigation systems, that they keep fertilizer to a minimum, and that land is cheap—the overall equation in Meta is favorable for upland rice.

As to CIAT's rice program, upland yields are high enough to warrant the research investment. Yields in favorable areas had increased from 1.5 metric tons in 1974–75 to almost 3 tons by the mid-1980s.[19] And Colombia is not the only country in which upland rice has potential. Neighboring Venezuela has its llanos too, which are also suitable for the crop. East of the Andes in the Department of Santa Cruz, Bolivia has fertile savannas for upland rice. And in Brazil, just south of the Amazon Basin in Tocantins State, conditions are also favorable. In fact, for South America overall, almost two-thirds of the rice acreage is planted in a rain-fed upland system. The research CIAT does on upland rice thus reflects the crucial role the crop plays in the region's agricultural development.

Santa Rosa

East of Villavicencio, at Santa Rosa in Meta, is CIAT's rice research station. I went there with Bob Zeigler, who wanted to be in Santa Rosa at planting time. It was mid-May, a good time to get field trials underway.[20]

In Villavicencio, CIAT has a guest house that doubles as a training center. Courses on rice production and pest control are held there, often in conjunction with ICA. In fact, CIAT's station at Santa Rosa and ICA's research center at La Libertad are just a couple of miles from each other.

The Santa Rosa Station gets plenty of rainfall for testing purposes—almost 2,500 millimeters. It rains so much the fields stay soggy without irrigation. Nonetheless, the soil is poor and has to be corrected by adding lime to reduce acidity and phosphorous and nitrogen to add to fertility. Meta is unique in that both irrigated and upland rice varieties are planted in such proximity. At Santa Rosa, for example, the trials are mostly for irrigated, modern varieties. Just 30 miles east, near the river town of Puerto

López, where soils are worse and rainfall less, CIAT conducts trials for upland rice.

"Many national programs do breeding independently," said Bob Zeigler. "What we stress at CIAT now is breeding for complex traits like blast resistance." The Santa Rosa Station is considered a blast "hot spot." Numerous strains of the wind-borne fungus are present and the infection rate for rice is high. Blast can strike at almost any point in the rice plant's life cycle, from seedling stage to panicle initiation. The fungus has so many ways to survive, it is almost impossible to destroy. The pathogen can lie dormant on seeds or on rice straw left in the field, and it lives in infected weed hosts all year around. That is why using fungicides is so often ineffective. Not only are the timing and method of application sight-specific, but new, resistant races quickly appear and farmers have to switch fungicides. Consequently, breeding resistant rice varieties seems the best way to control the fungus. [21]

Although blast disease is significant worldwide, it is particularly virulent in Latin America. Varieties that pass muster in Asia often succumb to the disease when tested at Santa Rosa. "Asian rices alleged to have blast resistance," said Zeigler, "can withstand only a few races or a low rate of infection. Most of them cannot survive the tough screening that occurs here; resistance breaks down quickly." When a line withstands blast disease at Santa Rosa, researchers can be confident the resistance will hold up elsewhere, and for much longer.

Rice is the world's most important food crop, which makes blast the world's top crop disease. "We do not understand all the dynamics of blast, such as how it shifts in the field, or the impact that climatic conditions and production practices have on its spread. The best we can do," Zeigler said, "is keep a step ahead of it." That is why a good testing place is so important. Otherwise, resistance breaks down quickly. To make sure the Santa Rosa lines become infected, trials are planted at high densities perpendicular to the prevailing winds. Susceptible varieties are planted around the plots. The weakest blast strains hit these varieties first. Only the hardiest, most virulent blast races make it through to infect the trials. "Because we are careful to keep blast infection levels up and our readings accurate," Zeigler said, "the lines selected here hold up well, even in Goiás State in Brazil, where blast is also severe."

Screening for blast resistance is an expensive, meticulous affair. Because the blast fungus mutates rapidly, resistance is unstable. So hundreds of rice varieties have to be infested, tracked, and evaluated annually. To make screening more efficient, CIAT has set up a joint research project with Purdue University. Using DNA "fingerprinting," the project is categorizing Colombia's highly variable blast fungus into fourteen different families. Prior screening had identified rice lines that withstand certain blast families. The objective now is to match tagged resistance genes in rice with particular blast families. With this information, scientists can breed resistant varieties quickly and farmers, in turn, can cut back on chemical controls.[22]

Seed Production

With its many squares and flower gardens, Villavicencio seems an attractive, unhurried place. Despite a population of 250,000, it did not have much traffic or urban sprawl. On the city's outskirts there are rice-processing mills and a seed-production plant.

For the Magdalena Valley and along the North Coast, FEDEARROZ manages seed production. In Meta, however, much of the region's seed came from Semillano, a private company. I talked with its director, Ernesto Andrade, a well-dressed, serious man, in a spacious, carpeted office.[23]

For 1990, the Department of Meta was expected to plant some 130,000 hectares of rice. Not all farmers purchased new seed each year; some planted from stocks held over from the previous crop. Nonetheless, most farmers did, which meant a demand for between 8 and 10 million kilos of seed. About half of this came from FEDEARROZ's stocks; the rest came from Semillano. "With public agencies in the seed business," Andrade said, "farmers face delays and end up with inferior seed. When budgets get cut, quality necessarily drops." According to Andrade, the keys to a successful seed program are volume, quality, and availability.

Semillano has its own research division. When ICA releases a new variety, Semillano does its own trials first before undertaking production. It is ICA's job to do random checkups to make sure seed quality is high enough for certification. In Semillano's case, such tests were perfunctory, given its good reputation. Each year, the company produces between 4 and 5 million kilos of rice seed on about 4,000 hectares. To avoid a buildup of insect pests and diseases, it does not use the same land for production twice in succession. It either shifts to new land or rotates to a leguminous crop, usually soybeans or cowpeas.

Semillano produces four different varieties of rice seed. The popular Oryzica, an upland rice suited to the llanos, accounted for about 60 percent of the company's sales. It also has seed trials underway for sunflowers, sorghum, and soybeans. "We want to have seed ready for crop rotations with rice," Andrade said. "We are sure more farmers will start rotating." After the rice harvest in late August, most farmers leave the land fallow. But they have time for another crop, as rainfall, though less, is dependable until December, just before the dry season. Semillano also works on hardy grass seed and forage legumes for a rice-pasture rotation.

Andrade gave me a tour of Semillano's installations. Prior to packaging, the rice seed is cleaned, air dried, and treated against diseases, notably blast. Once in storage, packaged seed has to be protected against insects and fungi, particularly during the rainy season when humidity is high. Seed is stacked over iron grates, placed so that fans underneath can keep the air circulating and the seed cool, deterring fungal growth. Semillano delivers the biggest orders directly to farmers in its own trucks and distributes seed to stores throughout Meta.

According to Andrade, without private companies there would not be enough high-quality seed available. Production has to be planned several seasons in advance. And for a new release, seed stocks must be built up prior to mass marketing, which is time-consuming and expensive. To keep a step ahead of blast, Semillano needs new varieties in the pipeline for when resistance breaks down.

Puerto López

Beyond Meta's piedmont belt, savanna soils take over. They are acidic, often with high concentrations of aluminum. Puerto López, only 50 miles east of Villavicencio, is a savanna town on the rice frontier. Farther east, soils are worse and prospects for upland rice dim. Despite abundant rainfall across the llanos as far east as the Orinoco River, it is poor agricultural country. Only the yellowish brown clumps of savanna grass seem to thrive. Each head of cattle needs several hectares just to survive. And although some upland rices tolerate acidic soils, yields are still low; around Puerto López, not much more than 1 ton per hectare.

An attractive little town, Puerto López is set back at some distance from the Metica River, which floods during the rainy season. Rustic wooden boats with old outboard motors are tied up to a makeshift dock, their outbound cargo mostly bottled beer, soft drinks, and mineral water, with or without "gas." The town has a quiet square of shade trees and flowers, with a church, a post office, and a branch of the Telecom telephone company.

Small shops and snack bars line both sides of the town's main street. Families take advantage of the cool night air, moving their chairs outside. From every cantina and radio comes salsa music.

At Hotel Tío Pepe, the rooms face an interior patio. Cars are parked inside, as important as the guests. Furnishings include an overhead fan, a dim light, two beds with lumpy mattresses, and a wooden chair. Like everyone else, we sat outside drinking beer, waiting for temperatures inside to cool down enough for sleep. I was with Bob Zeigler and Surapong Sarkarung, a breeder who worked on better upland rices for the savannas.[24] A hard-working, wiry man in denims and a t-shirt, Sarkarung seemed more at ease here than back at headquarters. Originally from Thailand, he had experience in Africa, where he had worked on rice at the International Institute for Tropical Agriculture (IITA) at Ibadan in Nigeria.

To breed for the llanos, the rice program crossed *Oryza sativa* upland rices from both Asia and Africa with irrigated varieties. "For high, stable yields," said Sarkarung, "an upland rice needs deep roots to withstand short dry spells, tolerance for highly acidic, aluminum saturated soils, and high yield potential." Because most upland rices are already tall, the nitrogen left over from a soybean rotation can weaken the stems. Improved upland rices from CIAT minimize such drawbacks. The savanna lines under test have high yields, good grain quality, and blast resistance, and they have strong stems than do not lodge and are deep rooting.

Between 1984 and 1989, a third of the genetic parents used in CIAT crosses were of African origin. African germ plasm helped to both improve the region's upland rice and diversify the genetic base of irrigated rice. Some of the African rices, for example, were upland semidwarfs, others had traits most irrigated varieties lacked, such as deep-rootedness and iron tolerance.[25]

The new upland lines were so superior that yields doubled with minimal inputs. "With a little nitrogen," Bob Zeigler said, "we get 3 metric tons per hectare. Even with zero inputs on poor soils we are getting 1.8 tons per hectare. That adds up to more than an incremental improvement, it is a major change in the yield potential for upland rice itself." Nonetheless, the problem is more complex than productivity. Most savanna farmers are also ranchers; they plant rice for a year or two and then rotate the land to pastures. Upland rice is just one element in the region's farming system, so higher yields alone do not solve the problem of depleted soils and degraded pastures.

In the llanos, given acidic soils and low-grade pastures, even a farm with 100 hectares is considered small. When farmers plant rice, they prepare the land by disc plowing, which breaks up the soil's surface to a depth of about 10 centimeters. To permit greater soil saturation and reduce erosion, however, deep plowing is far superior. When a chisel plow is used, which does a much better job of breaking up the soil, the rice plant's root system can penetrate to a depth of 25 centimeters. Once the soil is prepared, farmers add lime, nitrogen, and phosphorous—how much they invest depends on expected yields. Afterward, when they rotate to another crop or to pastures, they use profits from the rice crop to cover the expenses incurred. So if rice yields go up without adding much to production costs, the surplus adds significantly to the rotation options farmers can consider.

According to Zeigler, "CIAT's pastures program has a grass-legume combination that maintains soil fertility much longer, thus reducing the deterioration of pasture lands. The problem is, farmers cannot afford the added costs for seed and soil preparation. If they plant the new upland rice lines first, they can make enough to invest in a good pasture. After four or five years, they can return to a soybean crop followed by rice." Such rotations are mutually beneficial, as they help build the soil's organic content. Before such strategies can be recommended to farmers, however, CIAT has to be confident they pay off. So it needed to test the upland lines on a scale large enough, in blocks of 20 to 30 hectares, to be representative, not just on a small demonstration plot. And the tests had to be performed with varying input levels and competing methods of pasture establishment.

We left Hotel Tío Pepe at six o'clock the next morning and headed across the Meta River east of Puerto López for one of Sarkarung's 30-hectare test sites. If the new lines prove successful, the rice frontier could extend another 60 miles east to Puerto Gaitan, near the confluence of the Meta and Manacacías Rivers. "The rice-pasture rotation," noted Sarkarung, "has to be tried out with low inputs on a scale that matches on-farm conditions. Savanna soils in the llanos are so acidic and so nutrient deficient that farmers must add lime and phosphorous. Otherwise, the new lines simply will not be productive enough to cover the cost of rotations." Without any inputs at all, yields are about 1.3 metric tons per hectare. With moderately low inputs per hectare, 300 kilos of lime to reduce acidity and 60 kilos of phosphorus, yields increase to about 2.5 metric tons. That is more than enough to cover the cost of a high-quality pasture.

What costs do farmers incur? Farmers without tractors either rent the equipment and do the work themselves or hire someone with a tractor to prepare the soil. One pass with the disc comes to 5,000 pesos per hectare—about $14 in 1990. A pass with the chisel costs the same, but it is most effective when done twice, which adds as much as 10,000 pesos to the cost. Then lime and phosphorous must be applied, followed by incorporation into the soil—another pass. The cost of seed depends on the density of the planting; 180 kilos per hectare adds an additional 36,000 pesos. Then comes planting and harvesting. Surapong estimated that the total expense involved equals what a farmer makes on a yield of 1.4 tons per hectare. A ton of rice sells for about 90,000 pesos, so the total (90,000 pesos × 1.4 tons) comes to about 126,000 pesos, or $340. With a yield of 2 tons, farmers gross 180,000 pesos, for a net profit of 30 percent—some 54,000 pesos. Consequently, a yield of 2.5 tons is more than enough, as it leaves farmers with a surplus they can invest.

In the llanos, the soil's fertility is too low to follow rice with rice. So farmers rotate either to pastures or to a different crop. One advantage of upland rice is that it makes rotations easier. With irrigated rice, the ground is soggy or the soil is compacted. According to Bob Zeigler, a high-quality balanced pasture improves the structure and organic content of the soil. So later, when farmers rotate back to rice, or to a different crop, the soil has actually improved. The impasse is that farmers are reluctant to borrow money for pasture improvement—it takes them several years to recover the investment. But with rice, they can pay the loan back in just four months. So with enough profit on the rice crop, farmers can solve the rotation problem.

Zeigler described one possible rotation model. The available land, say, 150 hectares, is divided into five parcels, in this case of 30 hectares each. The first year, the farmer starts with rice in parcel one. The second year, he plants an improved pasture in parcel one and the rice crop in parcel two. In year three, he ends up with improved pastures in parcels one and two and rice in parcel three. At the end of the five-year cycle, parcel one goes back to rice. There are other variants. Farmers can start with six parcels, working in a soybean crop before rotating back to rice at the end of the cycle. Testing such variants is time-consuming but essential. The technology recommended to farmers has to prove itself first. Balancing inputs to maximize yields and minimize costs is the kind of calculation that makes the difference between success and failure. Pastures, for example, can be established with the residual fertility left over after the rice crop. Farmers do not need to add nitrogen. However, does it pay to add at least a little nitrogen? And when should the pasture be established? After the rice or along with it? The conventional approach is to stagger the rotations—rice first and then a pasture. This means farmers have to prepare the soil twice. But what if both rice and pasture can be planted simultaneously?

José Ignacio Sanz, a Colombian soil scientist at CIAT, explained that when rice, grasses, and legumes are planted together, the rice surges up first. "Farmers have to prepare the land only once," he said. "Inputs like lime and nitrogen benefit both, so overall costs are less. The rice is planted in rows and the mixed legume-grass pasture is broadcast over the field. When the rice is harvested, the loan farmers pay off covers both. They end up with a good rice harvest and the makings of a high-quality pasture, all in one season. Improved pastures planted separately are too expensive for most farmers."

It is simultaneous planting that farmers later adopted. According to a recent CIAT press release, both ranchers and crop farmers prefer it. On-farm trials show it is profitable for producers and that it protects savanna soils.[26]

TROPICAL PASTURES

Given the contribution rice makes to food production, investing in the crop's welfare is easy to justify. An improved pasture, by contrast, adds only indirectly to local food stocks through meat and milk production. Research by CIAT on pastures is focused regionally on Latin America. The rationale behind its work has as much to do with ecology as it does with food production. In the llanos of Colombia, in Brazil's cerrado, and in Bolivia's *chaco*, sustainable pastures and upland rice are linked together. So understanding CIAT's work on pastures helps explain why the two programs ended up working together.

Since the 1960s, cattle production in the tropics of Central and South America has proliferated enormously. So too has the land's degradation. What makes pasture research important is its focus on reclamation, its objective of helping farmers restore the land they have instead of clearing new acreage. In Colombia's llanos, farmers already rotate from rice to grazing. The challenge is to do this in a sustainable fashion, to improve pastures, increase milk production, and rotate crops so that the environment benefits. The task is urgent: the cattle frontier has already intruded into the Amazon, putting the rain forest at risk.[27]

A pernicious cycle impels the intrusion. To clear land, whether the grassy llanos or the stumpy trees and bush of Brazil's cerrado, farmers first burn over their acreage. Then meager subsistence crops—cassava, upland rice, and beans—are planted. The land is soon exhausted and low-grade pastures take over. The pastures, in turn, become so degraded and the land so eroded that farmers have to clear additional acreage. And so begins another round of burning. As a result, the cattle frontier has made its way to the Amazon Basin with disastrous results.

In Brazil, public policy reinforced the cycle. To prove occupancy on the frontier—as opposed to mere land speculation—claimants had to demonstrate their land was being productively used. The cheapest way to do this was to burn over what they claimed and graze a few head of cattle. In 1988, the Brazilian government eliminated such a blatant incentive to the forest's destruction. This alone, however, will not rewrite the equation that leads to beef.[28]

José Toledo, agronomist and head of CIAT's pasture's program, told me that tropical America has more than 250 million head of cattle—the largest tropical herd anywhere in the world.[29] Per capita beef consumption, not counting Argentina and Chile, exceeds 36 kilos a year—compared to 9 kilos in Africa and 3 kilos in Southeast Asia.

A pragmatic man, Toledo noted that CIAT's pasture research has its critics. They worry that better pastures will only speed up the pace of settlement. And they dislike ranching and beef production on principle. In Toledo's view, making pastures sustainable is a first step toward stabilizing tropical agriculture. And CIAT does not promote beef production. For Latin Americans, beef is part of the diet, and not just for the rich. Household surveys show that even poor families spend between 15 and 25 percent of their weekly food budget on meat. When the region's labor force shifted from the countryside to the cities, an expanding urban population created a demand for meat and milk that grew faster than the supply. So prices keep pace with inflation. "Farmers consider cattle raising a good investment," Toledo noted. "They never lose money on it. And most production is dual purpose, for both meat and milk, which creates flexibility. Farmers stick to milk production when meat prices are low; when they go up, they can sell off a few head for extra money." Settlement policies, poor soils, consumer demand, and the benefits cattle production brings to farmers creates a situation of short-term gains followed by long-term, irreversible loses. The problem is how to break the cycle.

Grasses and Legumes

Finding a suitable hardy grass for pastures was difficult. In the llanos, native grasslands have evolved without grazing pressure, as South America's great plains never had herds of grass-eating ruminants. Consequently, CIAT had to collect grasses from other regions of the world and then screen them for grazing, hardiness, acidic soils, and drought tolerance. "Germ plasm has to be tested first on small plots," noted Myles Fisher, a tall, thin Australian ecophysiologist who works with CIAT's pastures program.[30] "We start by analyzing a plant's growth pattern. Grasses start out slowly and then reach a peak. Since dead matter is replaced by new growth, they are self-sustaining. In pastures, the rates of senescence and regeneration are crucial. When combined with leaf density and overall biomass production, we have some basic criteria for selection. The next step is to evaluate the grasses selected under grazing conditions."

Tropical grasses have high growth rates but need at least some nitrogen to stimulate growth. In the llanos and the cerrado, however, soils are notoriously acidic and nitrogen deficient; growth drops off quickly. Most farmers cannot afford chemical fertilizers just for pastures. The solution is an organic combination of nitrogen-fixing legumes and grasses. The trick is to find the right balance.

"In the tropics, a sustainable pasture needs a legume component," Fisher said, "but grasses tend to grow faster. In fact, the grass will dominate unless the legume has an advantage that compensates, such as greater drought tolerance. Such factors have to be evaluated under grazing. As a forage, cattle prefer grass to legumes. They will stick with the grass all during the rainy season, a period of vigorous growth. When the dry season comes, the grass gives out first. Legumes, however, have a deeper root system. They stay green much longer and provide forage during the dry season. So what we have to do is balance rapidly growing grasses with legumes that can compete." Legumes interspersed with grasses is the most popular approach, sometimes in association with separate dry-season legume parcels held in reserve.

A low-grade pasture means less milk production, weight loss in cattle, and low rates of animal reproduction. Thus, a successful program must show farmers that the added work and expense of improved pastures pays off in more milk, more calves, and more beef.

Testing

It took almost twenty years for CIAT's research on pastures to pay off. Why it took so long reflects the novelty of the project's agenda. Rice has been cultivated for thousands of years. By contrast, research on tropical forage grasses has few precedents; CIAT's work, begun in 1970, virtually started from scratch. And it faced serious ecological constraints, from poor soils to long dry spells. It took a decade to select promising grasses and run them through the requisite trials. Progress was slow, particularly given the problem's multifaceted aspects.

Patricia Avila, a research assistant who has worked on pasture quality for more than five years, took me to CIAT's testing site at Quilichao, in the foothills southeast of Cali.[31] Short, thin, and articulate, Avila was a graduate of Colombia's Agrarian University at Palmira, not far from CIAT. She explained why the research took so long. The main factors in testing, Avila noted, are the soil, the grass and legumes, the animals, and the system's overall production. Trials also must consider the climate, various pests and diseases, and pasture management, that is, the interval for grazing as opposed to leaving the pasture at rest. "The traditional research system was rigid," Avila said. "Its objective was to evaluate grasses and legumes under fixed grazing levels. We wanted to know how well different legumes did during the dry season. To what extent did they stay green and provide forage?" Standard research used a hectare divided into four parcels. Each parcel was grazed with cattle for seven days and then rested for twenty-one days. Researchers assessed how well the legume survived. Three treatments were considered: heavy grazing—four head, medium grazing—three head, or low grazing—two head. The percentage of the pasture sown to legumes varied from 15 percent to half. So the factors considered were variations in the percentage of grasses and legumes, the number of cattle grazed, and the period devoted to rest and use.

According to Avila, the problem with the fixed method was that it took so much time to conclude anything. So researchers designed a more flexible approach that looked at the management of the pasture system as a whole. From previous research they had a good idea of what the optimal grazing level was and the kind of grass-legume association that was viable. The problem was determining the right balance. When a unit was being overgrazed they took animals off; at other times, they added. "In this way, we could test different combinations and reach conclusions faster," she said. "We now have production strategies we can recommend to farmers."

At the Quilichao Station, the parcels tested had different grass-legume combinations subject to varying grazing levels. *Andropogon gayanus*, a hardy Nigerian grass with thick, wide blades, grows in dense clumps. Uncut, it reaches a height of well over a meter; from a distance it looks more like sugarcane than pasture grass.[32] A nutritious plant with high protein content, it is a good for calves, whose nutrient requirements are high. And it does well mixed with legumes. During the dry season, however, *Andropogon* fares poorly, making it impractical on a large scale. For overall hardiness, Avila thought the best choice was the East African grass *Brachiaria dictyoneura*. It combines well with two vigorous, digestible legumes: *Centrosema acutifolium*, which is native to the llanos and known as "vichada," and *Desmodium ovalifolium*, from Southeast Asia.[33] Such selections had already passed susceptibility tests for many insect pests and diseases. *Brachiaria dictyoneura*, for example, resists the notorious spittlebug. It also helps shore up denuded soils against erosion.

In Mondomo County, which borders the Quilichao test site, soils are so poor they have to be rested for six to eight years, even after planting cassava, a crop that tolerates marginal soils. For any given year, about 70 percent of Mondomo's land cannot be cultivated. Raúl Botero, a pastures agronomist from Colombia, took me to see the project.[34] A practical man with much experience, Botero based his remarks on farmers and local practices. "The objective is the soil's recuperation," he said. "After a crop, farmers plant a hardy pasture that inhibits erosion—instead of just leaving the land fallow. The soil usually has enough nitrogen and organic matter left to get the pastures started. Later on, legumes provide the nitrogen." We left the jeep by the road and hiked up to parcels where farmers had trials underway. The best combination was *Brachiaria dictyoneura* in conjunction with the vichada legume. In just a month, the robust, crawling dictyoneura had put out runners that had rooted. After nine months the pasture was ready for grazing.

Farmers liked the dictyoneura because it grows so fast. To plant the grass, farmers use rhizome cuttings, which the project provides at fifty pesos a kilo—about fifteen cents. With an improved pasture, the grazing load can be increased to several head per hectare instead of just one. Farmers keep cattle for both meat and milk. Marketing is not a problem because Mondomo is close to Cali and roads are good. "Farmers can already see how good the dictyoneura is," said Botero. "Of course, we had to do trials first to make sure the dictyoneura did well, even without fertilizer."

"The ultimate objective," Botero said, "is a more balanced approach to agriculture. No single crop can accomplish this by itself. In fact, the traditional way of farming recognizes this; it is much more diversified." Botero pointed out how well local farms make use of shade trees, which protect coffee bushes and cacao trees. Many farms also have vegetable gardens and citrus groves. "We need to combine pastures with forestry," he emphasized. "Rapidly growing timber and robust shrubs can be planted around fields for firewood. Upright trees with small leaves filter the sunlight, and hence, do not overshadow the pasture. They make good fences too, as wire can be strung from tree to tree."

Components

It was some time before I realized a pasture was just one aspect of a farm. To me, cattle meant ranches, crops meant farms, fruit trees meant orchards, and timber meant forests. For sustainable agriculture in the tropics, such compartmentalized thinking has to be set aside. In the llanos, rice and pastures are linked in rotations that can be mutually advantageous or self-destructive. Not even pastures and trees are antithetical. When trees are properly spaced and a suitable timber species is used, they can be planted with pastures.

"Research on a forestry component for pastures is just starting," noted Rainer Schultze-Kraft, a West German who worked on testing new germ plasm.[35] "What we want are trees that permit grazing underneath, that do not inhibit a pasture's formation and growth." Algaroba and acacia are good examples. They grow rapidly and are leguminous, which helps the soil's fertility. They can be planted as a fence and then cut back for firewood. The leaves make good animal fodder. Leucaena, which is native to Central America, does well in poor soils, even in areas with prolonged dry spells. For the rainy tropics, the leguminous poro tree is a good choice. In Costa Rica's highlands, where soils are rich and deep, farmers plant it around their fields to make a "living" fence. Poor, acidic soils, however, like those at the Quilichao Station, stunt its growth. The African palm is suitable for a multipurpose farming system. Its abundant seeds can be harvested commercially for cooking oil. Because it is not a shade tree, crops can be planted underneath or the land rotated to pastures for grazing. Shrubs also play a key role. Flemingia, from Southeast Asia, makes a good green manure; calliandra, from Java, has nitrogen-fixing nodules, so it does well in poor soils. Both provide forage and firewood.[36]

The environments for cattle production are diverse, as are CIAT's testing sites. The Ecuadorian site at Napo is in the high rainfall tropics of the Andean foothills. Compared to Meta in Colombia, Napo does not have a long dry season. According to Raúl Vera, an animal scientist at CIAT, Napo is a region of small-scale production.[37] Ecuador's land-reform agency had divided the land into parcels, most of which did not exceed 10 hectares. "Some land is put into pastures," Vera said, "some is used for crop production, and some is planted with fruit trees and timber. In collaboration with Ecuador's Agrarian Institute, we try to combine land uses in a sustainable way."

Cooperative, interdisciplinary research characterized other sites too. Near Pucallpa, a small Amazonian town in Peru on the Ucayali River, research combined agroforestry, anthropology, and agricultural economics. CIAT also had outposted staff at CPAC, Brazil's cerrado center outside Brasília. For Colombia, the main testing site was at Carimagua, a frontier site in the savannas beyond Puerto Gaitan.

"We do plenty of research," said Carlos Seré, an agricultural economist from Uruguay. He was more than six feet tall, mustached, and impatient. "We need less research and more action. To see if a dog bites, give it a good kick. It's time to kick."[38] In conjunction with the ICA, Seré worked on a demonstration project in Caquetá. The project involved extension workers, dairy cooperatives, and banks. "We need linkages with local institutions and with farmers," Carlos said, "otherwise, our work will seem like another intrusion by outsiders." Carlos had a trip to Caquetá scheduled with Raúl Botero for later in the week and invited me to go along.

CAQUETÁ

Seré and I left Cali for Florencia, capital of Caquetá Department, at 5:00 A.M. The trip covers a distance of more than 300 miles and crosses two mountain ranges. The first passage was across the austere altiplano of the Cordillera Central and down into the lush southern reaches of the Magdalena Valley, rich in cotton, sorghum, and tropical fruits. Then, once through the valley, we started up the slopes of the Cordillera Oriental. Having crossed the divide, we began the descent. Torrential rains soon set in, turning the compacted dirt road to mud. Water rushed down the mountain sides, crashing against the rocks and washing out the road. Oncoming traffic consisted mostly of overloaded cattle trucks that tilted dangerously in a maze of muddy switchbacks. We kept going down, for there was no going back.

Florencia is situated in the rainy highland tropics at an altitude of about 500 meters and has about 100,000 inhabitants. Compared to the sleepy provincial towns of the Magdalena Valley, it is a dynamic place of expectation and modernity. Here, on the edge of the Amazon basin, shops bulge with the electrified gadgetry of the twenty-first century: from Walkmans and ghetto blasters to VCRs and PCs. Flown in as an illegal byproduct of the drug trade, such untaxed goods were cheaper here than in Bogotá or Cali. The edge of the Amazon is ringed by Florencias: Iquitos and Pucallpa in Peru, Santa Cruz in Bolivia, Rio Branco, Porto Velho, and Imperiatriz in Brazil. It will not be easy to tame the forces arrayed against the Amazon's fragile ecology.

Caquetá takes its name from the river that marks the province's southern boundary. Rising in the Cordillera Oriental, the Caquetá flows southeast for more than 2,000 miles before joining the Amazon, deep within Brazilian territory. Caquetá Province comprises some 34,000 square miles; it is almost the size of the state of Indiana. Settlements, however, cling to a narrow north-south strip, the "high tropics" of the Andean foothills. In Caquetá, this fertile belt of rolling hills and flatlands covers some 120 miles north to south but rarely exceeds 40 miles east to west. From Florencia, a single paved road, financed by the World Bank, extends north and south, reinforcing the city's commercial position. A gently sloping terrain made the road easy to pave and maintain. The main hazard is fluvial. Streams from the foothills cut across the road everywhere, heading for the main rivers. And each stream needs a sturdy bridge to withstand rainy-season floods. Dirt and gravel roads spur off to the east and west, connecting farms and towns to the main road. Beyond the foothills the lowland tropics begin, a marshy, inaccessible world of rivers, forests, and cocaine that is almost impossible to patrol.

Mountain streams have deposited rich organic matter in the valleys that skirt the eastern Andes. It is in this north-south strip that agriculture is well established, particularly dairy farming. Around Florencia, for example, the phosphorous level is twice that of Colombia's llanos or Brazil's cerrado. And the dry season, which lasts six months in the llanos, is much reduced. In the Caquetá foothills, it does not begin until December and is over by the end of March—just four months. And sporadic downpours, even during the dry season, are not uncommon.

Settlement

Caquetá had a rubber boom before settlers began cutting back and burning the forests in the 1950s.[39] Then came the drug trade, first marijuana and later cocaine. "Through all this turbulence," Seré said, "it was meat and milk production that most settlers depended on, and which made a livelihood in this part of the tropics possible."

The World Bank provided loans to help Colombia carry out a land-distribution program. Most of the region's twelve hundred dairy farms and ranches cannot be considered agribusiness enterprises. Nonetheless, compared to how Colombia's Rural Development Agency (DRI) defines a small land holding, which is 20 hectares or less, farms in the Caquetá region are much larger. But they have to be, because a pasture's carrying capacity rarely exceeds one head of cattle per hectare. "Around Florencia," Raúl Botero said, "a farmer needs at least 40 hectares to support a family. And even 100 hectares is still small for a dairy farm. Most of the ranches, even medium-sized ones with 300 to 400 hectares, are not large in terms of family income. After all, the question is how much land is suitable for pastures. Farmers clear the flatlands first, they do not deforest a mountainous terrain unless they run out of better land. Someone with 400 hectares of land may have only 150 in pastures."

There are many reasons why Caquetá's farmers keep cattle.[40] Once cleared, the soil's fertility declines rapidly. Without costly inputs of nitrogen fertilizer, crop production cannot be sustained for long. So farmers quickly convert the land to pastures. Alternatives that seem viable on paper, such as coffee and citrus trees, face almost insurmountable obstacles. Colombia already has a coffee surplus; hence, adding production in Caquetá is not profitable. In fact, Colombia's peculiar geography is in part the culprit. Given three north-south mountain ranges and deep valleys like the Magdalena, almost every department in the country has a tropical zone for exotic fruits and citrus or a cooler zone for coffee. So for agricultural production, Colombia hardly needs Caquetá at all. One of the few products that can be carted across the mountains profitably is cattle on the hoof and milk, mainly because of the growing demand in big cities such as Cali. And cattle production does not have high entry costs: with their own meager resources as security, farmers can clear some land, plant a crop, and convert to pastures.

Around Florencia, the tropical surroundings seem luxuriant, even though the forest was sacrificed to pastures. The hills stretch gently to the distant mountains; patches of rain in the distance set off an azure-blue sky with great billowy clouds. The cool air is refreshing. Such is the panoramic effect. Close up, a forest in the midst of destruction is a terrible sight: charred trees, debris, defoliation, a wasteland. When cleared, farmers get a stand of shrivelled, diseased corn and a cassava crop. Then they convert to pastures. When the pasture gives out, more land is cleared. To break the cycle, intensifying agriculture so as to slow the forest's destruction is a necessary first step. With better pastures, farmers can get more milk and feed more cattle.

A sustainable pasture reduces the pressure to clear marginal, hillside land. The challenge is to get farmers to try the technology.

Nestlé

To convince farmers to spend more on pastures requires an assertive strategy.[41] The ICA regional office at Macagual outside Florencia provides local expertise. Also important is technical support from the milk processing business, particularly the local Nestlé subsidiary. The company has a good reputation; in fact, its impact on the region's development is substantial.

As soon as hydroelectric power was available and a paved road made local access reliable, the Nestlé Company set up its first milk processing plant. Nestlé provides an alternative to farmers: they can sell milk to Nestlé instead of cattle on the hoof to truckers. The company pays less for raw milk in a frontier zone such as Caquetá, but the outlet it provides is still beneficial to farmers. "Nestlé picks up the milk on schedule and always pays on time," noted Jiro Gomez, who works with the Agrarian Institute's local training program.[42] Jiro is a short, tough man with dark, weather-beaten skin; we had lunch together at the Macagual Station. "Nestlé checks circulate like cash, making up for the lack of banks. By the time a check is cashed or deposited, it has six to ten signatures on it. Farmers have confidence in the company; they consider it a serious business." Local ranching quickly became dual purpose: farmers switched back and forth from milk to meat sales, depending on market prices. "When a cow's yield falls below a liter (3.78 liters = 1 gallon) a day," said Gomez, "farmers start to consider the alternative—the meat market." But this depends on how a farmer calculates. At a liter per day, a farmer has to consider how much that adds up to in six months, or in a year. The issue is overall income generation. "Right now," said Gomez, "unless a farmer needs the cash, it is better to keep the cow."

Nestlé is no longer the sole milk purchaser in Caquetá, but it is still the largest. To keep its facilities operating at capacity, it needs a dependable milk supply, so the company policy is to promote dairy farming. For example, it multiplies pasture rootstock for distribution to local producers and imports bulls to help improve local herds. In daily contact with farmers, Nestlé knows more about local production than CIAT or ICA. Consequently, it is important to enlist its support. "I began by inviting Nestlé personnel to see the results of pasture trials at Mondomo," Carlos Seré told me. "They were impressed and agreed to support the Caquetá project. We need farmers willing to work with us. Nestlé has local knowledge; it knows which farmers are dependable and trustworthy."

Doncello

We spent five days in Caquetá stopping at Nestlé plants, local farms, and extension stations.[43] The project's aim is sustainable pastures, better erosion control, and additional milk production. To spread the new technology, it has to show success locally. To do that, it has to work directly with farmers. We started out at Nestlé's Doncello plant, about 30 miles northeast of Floriencia. The town is set in a broad valley flanked by mountains.

Jorge Roza, the plant manager, took us through the Doncello facility. It was spotlessly clean and efficiently run. To detect changes in local production, it keeps track of the volume, by weight, from each milk route. At the Doncello plant, the milk is cooled to between 35 and 37 degrees Fahrenheit and pumped into thermal trucks. Even though not refrigerated, a truck's long cylindrical storage tank keeps the milk cool all the way to Cali. On arrival, its temperature rarely exceeds 40 degrees. Nestlé's main plant in Florencia has extra storage capacity, so even when the road across the mountains is closed, the backlog can be accommodated. And the road, which has a permanent maintenance crew, rarely shuts down for more than a day.

Pasture Establishment

In fields near its plants in Florencia, Doncello, and San Vincente, Nestlé tries out new pasture technology. Farmers stop by regularly to take a look.

How is a new pasture established? In Caquetá, farmers usually plant rootstock, not grass seed. They space out the cuttings in a field that has just been grazed. When the new grass is a *Brachiaria* and the old one a local variety, the more aggressive *Brachiaria* proliferates. How densely cuttings are planted determines how long it takes the *Brachiaria* to attain dominance. Sometimes farmers clear out the old grasses and weeds first. When this is done, the *Brachiaria* gets established even faster, as it has less competition. In the llanos, farmers burn over their fields in the dry season. In Caquetá, which has showers intermittently during the dry season, fields retain more moisture. So although burning can destroy surface vegetation, it does not always kill off the roots. The result is a new crop of weeds that comes back with renewed vigor. Many farmers end up applying a herbicide.

"Burning over a field," Raúl Botero said, "is the traditional way. It does not cost anything, and it has some positive results; for example, it releases nutrients into the soil from plant residues." How a farmer clears a field depends on how much he can afford to spend. "If he applies herbicides," Botero noted, "and then plows the residues under, he gets the best results. But it also costs the most."

Farmers establish new pastures in April and May, when the rainy season begins. Our trip to Caquetá coincided with pasture planting; wherever we went, people were out in their fields. The first step is to bundle up the cuttings from a *Brachiaria* plot. Next, the cutting are carted to the new field, where they are planted in a pattern. Spacing determines how fast a pasture can be ready for grazing. The requisite technology is simple: a sturdy pointed pole to make a hole in the soggy ground. A clump of grass is inserted in the hole and then firmed up by stepping on it. On some farms, the family itself provides all the labor. On others, farmers hire help locally.

Caquetá's farmer-ranchers appreciate a good pasture. "What makes improvements attractive," Carlos Seré said, "is that the extra cash needed is minimal. Family labor often suffices and cuttings can be obtained at a very low cost. For farmers, improving a pasture is a way for them to capitalize their labor. The return comes with greater weight gain in cattle and better milk production."

Greater milk production is not just a function of pasture quality and animal nutrition. It also depends on a breed's genetic potential. The Zebu, for example, fares well in the tropics. In Caquetá, even when grazed exclusively on traditional pastures, a Zebu averages 3 liters a day. With improved pastures, production doubles. More common, however, is the mixed criollo stock. It is a hardy breed but a poor milk producer—about 1 liter a day. Nonetheless, when grazed on improved pastures, criollo production rises to about 2.5 liters. Given the gain in milk production, an improved pasture quickly pays for itself.

The most popular pasture grass in Caquetá was *Brachiaria decumbens*, an African variety introduced into Colombia from Brazil. An aggressive, nutritious grass, it is resistant to the *cucurón*, one of the region's worst insect pests. And decumbens recuperates rapidly when the rainy season begins. Promoted by Nestlé, it spread quickly but has drawbacks. It is not a sweet grass and can be toxic to cattle, particularly calves.[44] Too much decumbens makes them photosensitive; consequently, they develop skin sores that fester and bleed. According to Botero, for every one hundred head, toxicity kills three or four a year and leads to serious weight loss for ten. Moreover, draft animals like horses will not eat it.

To walk through a pasture in the rainy season, even through a hardy decumbens pasture, is to end up ankle deep in a soggy mixture of grass and mud. When the rains are particularly severe, cows simply stop eating, loose weight, and milk production drops. Cattle killed off during the rainy season often have as much mud in their stomachs as grass. A thicker, clumpier grass like decumbens provides a better ground cover and reduces mud ingestion. Unfortunately, decumbens falls prey to the spittlebug.

The spittlebug is about a quarter of an inch long. They lay their eggs in a protective glob that looks like beaten egg white just before it stiffens. When I walked through an infested field, my boots got covered with it. Once the larva hatch, they feed on the grass, which then turns yellow. A severe infestation kills the grass.

Options

Using the fields that adjoined the Nestlé plants, and in conjunction with local farmers, the project evaluated how different grass-legume combinations held up in Caquetá and compared different planting methods, which included plowing and herbicide applications.

To get off to a good start, legumes should be planted first. Otherwise, a vigorous grass will crowd them out. Once established, a balanced pasture is sustainable for many years. Grazing cattle favor grass during the rainy season. In the dry season, when the grass turns brown, they switch to the legumes.

For Caquetá, the *Brachiaria* grass dictyoneura, which CIAT tested at Mondomo, has several advantages over decumbens. An aggressive, thick-bladed grass that spreads rapidly and clumps at its roots, it makes for a tougher ground cover than decumbens. Dictyoneura does well in poorly drained soils too, a circumstance quite common in the valleys. Horses like it because it is sweet. Moreover, dictyoneura is nontoxic to cattle and resistant to both the *cucurón* and the spittlebug. Another good choice is *Brachiaria humidicola*, which is even more vigorous and aggressive. It does exceptionally well in poorly drained soils, withstanding inundations better than dictyoneura. Humidicola thrives even in the phosphorous deficient soils of the llanos and cerrado. Brazil has some 6 million hectares in humidicola. It is good for eroded soils, where rapid establishment and vigorous growth are essential. It also helps the environment. A hectare of deep-rooted *Brachiaria humidicola* absorbs 53 tons of carbon dioxide from the atmosphere and converts it to organic matter.[45]

Planting legumes along with pasture grass is a new technology; few farmers in the region have tried it. As was true at the Quilichao Station, the legumes centrosema and desmodium grow fast enough in Caquetá to compete with the brachiaria grasses. *Arachis pintoi*, or wild peanut, also does well. A good leguminous forage, it roots quickly from crawlers that grow up and around the grass. Its ancillary shoots also make good cuttings for transplanting. When cut back, a field of *Arachis* quickly resprouts.[46]

The project did not promote any particular grass, legume, or combination thereof. Instead, farmers were advised to keep the genetic mix in their fields as diverse as possible. Different combinations in different pastures help protect against insect infestations and diseases.

I had a chance to check out demonstration plots on fields at both Nestlé and ICA. There were various mixes of legumes—*Centrosema*, *Desmodium*, *Arachis*—with the *Brachiarias* dictyoneura and humidicola. On plots left to grow ungrazed for a year, the resulting pasture seemed almost indestructible. The growth was waist high and so thick my boots did not even get muddy. A dense mix of crawlers and offshoots crisscrossed everywhere; it was a struggle just to walk.

Farms

"When farmers clear the forest and burn the debris," Carlos Seré said, "the soil ends up with enough nutrients to establish a good pasture.[47] After a few years, however, given the soil's lack of nitrogen, the pasture deteriorates. The carrying capacity for grazing drops from about two head per hectare to less than one. So as the pastures give out, the pressure to clear more land increases." With old pastures, the project wants to restore them before depletion gets out of hand. With new pastures, the objective is to start off with a sustainable balance of aggressive grasses and hardy legumes.

Most farmers are eager to try out the new technology. The dry season always takes its toll on pastures: the grass stops growing and overgrazing does a lot of damage. Farmers burn off old pastures so they can replant them when the rainy season starts, but they have to work fast before the weeds get a head start. Each cycle depletes the soil further. The prospect of halting such decline evoked great interest.

We visited farmers to arrange local trials with them. The project promises enough grass and legume cuttings to cover at least 3 hectares, along with herbicides to kill off the weeds. Farmers agree to plant the pastures and fence off the improved parcels. By grazing cattle on them exclusively for fixed periods, it is easy to measure how much milk production goes up. Why do farmers want to participate? There is the prospect of free cuttings and better pastures. But even more important is the prospect of good advice on a regular basis.

Señor Carrillo had on jeans, a short-sleeved shirt, a broad-rimmed straw hat, and rubber boots. Given the rain and mud, the boots were standard; the straw hat was optional. He was about fifty years old, short and stocky. He had a total of 220 hectares, he told us, of which some 120 were arable. Half of that was in pastures, of which 45 hectares were in decumbens. Carlos Seré described Carrillo as an intensely curious man, an early experimenter with new techniques. Carrillo's wooden frame farmhouse had a tiled roof and a comfortable veranda on three sides. Rainwater from the roof collected in a cistern, from whence it was pumped into an overhead tank and then gravity fed into the house. The house was cool and shaded by fruit trees—limes, coconuts, and bananas. There was a garden, too, with root crops of cassava and elephant-eared taro.

Carrillo took us to inspect his fields and cattle. He had planted some parcels of dictyoneura with legumes mixed in, all planted by insertion. "Every thirty days," he said, "after the parcels are rested from grazing, the dictyoneura grass recovers quickly. The *Centrosema* legume, trampled on in grazing, also recovers well." He had a field in humidicola too, a grass he especially liked because it "grew so fast," spreading out and "rooting in a circular fashion." What is the most effective way to establish a pasture? Most of the time Carrillo burns over parcels first, but sometimes he uses herbicides. He feels that burning is better, a conclusion reinforced when one of his workers applied so strong a herbicide mixture that it killed everything. "Not even the cuttings took root," he said. In general, it takes four months to get a good pasture, he said, estimating the labor involved as about thirteen days of work. Considering both labor and materials, it cost about one hundred dollars to establish a hectare of new pasture.

We rested in hammocks on Carrillo's veranda, discussing calves, weaning, and weights. After lunch, which included fresh coconuts, we continued our sojourn. Taking dusty dirt roads we headed for Don Pedro's farm, which is set back in an isolated, mountain-flanked valley. Don Pedro is thirty-five and has five children. Two years ago, he purchased 118 hectares to start a farm. He saved the money by working as a trucker, buying and selling coffee.

His cement-block house with a porch in front was almost finished. Don Pedro told us he was switching over from native grasses to decumbens; he had planted 18 hectares of new pastures and milk production had increased substantially. He had no prior experience with legumes but agreed to mix them in as necessary. Like many farmers in the region, he knew from experience that decumbens had trouble with spittlebugs.

We chatted about milk prices and decumbens toxicity in calves. But it was already late afternoon and we still had more farms on our list. By the time we made our last stop at Doña Antonia's, it was dusk. She has 40 hectares in pasture grass and thirty cows. Doña Antonia is forty, "more or less," and since her husband died, she has run the farm herself and cared for her five children. Her comfortable wooden house has freshly painted, bright blue, vertical clapboards. The veranda is in the back, facing a courtyard filled with flowers and fruit trees.

She invited us in for hot black coffee and passion fruit then took us to see one of her pastures. It was not often so many experts stopped by her farm; she was not going to pass up a chance for advice. She led the way at a rapid pace; we had to run to keep up with her. We were all equipped with rugged, rubber boots, but Doña Antonia managed well enough with a pair of sandals. She had burned over her pastures, she told us, but all she got back was "an incredible quantity of weeds, worse than before." Next we saw her fields of upland rice, and then her sick cow. We had veterinarians in the group, and they lined up to display their knowledge. The diagnosis: "the cow is sick." The explanation: "this happens to cows." "They do fine one day," it was said, "and then the next they stop giving milk." "It's probably cancer," someone suggested, "whose effects show up suddenly. But now it's too late, better send the cow off to the meat market." Doña Antonia seemed in no mood to lose a cow. "How easy to give advice," she said, "when the cow isn't yours."

DETOURS

It would seem that farms in Caquetá are far off the rice track. For success in upland rice, however, the research CIAT performed on pastures there turned out to be almost as important as its work in rice.

In 1991, ICA released Oryzica Sabana 6, a new upland variety selected by Dr. Surapong Sarkarung. In recognition of his work, Sarkarung received CIAT's distinguished research award.[48] Sabana 6 is the first high-yielding, deep-rooted upland rice to tolerate acid soils. It matures rapidly, in less than 110 days, and is blast resistant. How the crop is established is remarkable, for

Sabana 6 is part of a rice-pasture cropping system. Farmers first prepare the land and fertilize it. Next, they dry-seed the rice in rows spaced about 34 centimeters apart. Between the rows they broadcast a mixture of improved grasses (including *Brachiaria dictyoneura*) and legumes (including *Arachis pintoi*) developed by CIAT.[49]

As of 1992, Colombian farmers had tried out the new system on 6,000 hectares in the llanos. Yields from the stiff-stemmed Sabana 6 averaged 3.1 metric tons per hectare. The rice alone covered the cost of the new cropping system. Three months after the rice harvest, farmers could use the fields for grazing. The hardy grass-legume mixture supported two head of cattle per hectare. The overall weight gain per head was 174 kilos a year compared to 95 kilos on native grasses; and herds increased much faster. Finally, when farmers rotate back to rice, the crop benefits form the nitrogen already fixed by pasture legumes and from better soil texture.[50]

Small holders in the llanos who could not grow rice profitably before are now doing so. The new system has also spread to Brazil, where farmers have used it to restore some 150,000 hectares of degraded pastures. The approach is just a first step toward sustainable agriculture in the savannas. For as CIAT scientist Raúl Vera noted recently, stable agropastoral systems need to combine many crop components. Vera was leader of a new CIAT program focused specifically on the savannas and resource management.[51]

In Asia, rice seems like the whole story. In Latin America, it is just a chapter that leads to other crops and production systems.

CONCLUSIONS

5

What is rice? The short answer: hard white grains packaged in cellophane sacks at the supermarket. The long answer: it depends on where the rice is grown and how it is produced. In most of Asia, rice is not a crop, it is a way of life. Grains come in diverse shapes and sizes, in different colors and textures. Rice can be pounded into flour for use in sweet, sticky cakes or it can end up as saki; there are aromatic rices for special occasions; and then, as in Manila, there is simply rice, rice, rice—for breakfast, lunch, and dinner. The many faces of rice began at Jolibee's for breakfast and ended up, far afield, on Doña Antonia's back porch. For in South America's savannas, a story about rice cannot be told without a subplot of cattle and pastures. Stopping there is arbitrary. A story about rice has no ending. That is why rice at the supermarket is so misleading.

ONE SIZE FITS ALL

In industrial nations, the fast track shapes how we think. Consider development. When I ask my students what it means, I get back a predictable list of highways, factories, computers, and VCRs—in short, technology. And that is right. We think of development as a standardized product; all a country needs is the money to buy it. Modern life provides us with a pattern. From our jeans and sneakers to the best-seller list and the low-fat diet, we are standardized products. We think along the same track too. For every problem, we expect a solution. We want social problems solved the same way we buy stretch socks: one size fits all. For railroad tracks and generic drugs, for telephone lines and electrical wiring, for high-density disks and software, there is nothing wrong with a standardized product. But what happens when parts are not interchangeable, when systems are incompatible, when every loaf baked requires a different set of ingredients? What happens when prepackaged solutions do not fit the problem, when every puzzle has a missing piece? What happens when "development" has to fit the constraints inherent in a thousand different localities and situations?

"If development is that complex," my students tell me, "then it is hopeless. There never will be a solution." Gearing production to local constraints, however, is precisely the problem faced in rice. Even though there is no single solution, the problem turns out to be far from hopeless. What the rice story illustrates is how important the right approach is, even in situations in which solutions can be temporary, in which today's answer is just a stepping stone to tomorrow's question. When it comes to a Saturn or a Classic Coke, an assembly-line mentality is just what we want. But many of our planet's problems, from energy use to family planning, are more like rice. With this in mind, let us review the approach people took to growing rice.

Simultaneous Equations

To understand the rice story, we analyzed plots one by one. For the sake of analysis, we had to create an illusion—that the problem rice poses could be broken down by chapter and paragraph into discreet elements. It is better, however, to think of rice production as a chain of concurrent events, as a set of simultaneous equations whose answer depends on solving for all the factors included. Rice does not reduce to the work done at IRRI or by a national research program. It is more than a story about technology transfer and local extension. To describe how rice is produced, wading through irrigated paddies is not enough. To understand the prospects for rice self-sufficiency in Asia, germ-plasm banks and biotechnology are only part of the story.

Growing enough rice depends on many things. This book has looked at rice across a range of ecologies, countries, and localities. In so doing, some factors held constant, some changed, some even dropped out of the equation altogether. At times new components had to be added. So rice is not a single equation at all but, rather, sets of equations whose factors constantly change. To become overly attached to any single element, to any specific link in the chain, can spell disaster. For rice production is a multifaceted, highly interdependent activity, a chain of events that moves back and forth from research projects to demonstration plots to a farmer's field. Everything happens at the same time.

Revolutions

The rice revolution began where it could have the greatest impact as quickly as possible. That meant varieties and production technology geared to monsoon Asia's irrigated rice belt. It was impossible to solve all the problems involved in rice production everywhere—and still is. The impact of IRRI began with IR8, a stiff-stemmed semidwarf indica rice whose bred-in genetic traits revolutionized tropical rice production. It did not lodge, it converted nitrogen fertilizer to grain more efficiently, and it matured one to two months earlier than traditional varieties. It was nonsensitive to daylength, so farmers could grow two or more crops per year. And it could grow at many latitudes. This improved rice plant did so much better that it had an independent, positive impact on yields. To get the most from the plant's genetic potential, however, required more nitrogen and more exacting water control in rice paddies. Backed up by Asia's rice-testing network, INGER, breeders continued to make improvements. Compared to IR8, later releases did better against diseases such as blast and against pest attacks, notably by plant hoppers. There are now indica rices with greater cold tolerance. Some can even withstand salt water intrusion.

The decisive factor in Asia's rice revolution was the genetically improved rice plant, at least at first. Over the long haul, the equation becomes more complex. Changes in production technology, from how the land is prepared to nitrogen insertion and water control, began to account for a greater share of the yield increase. The precise balance among production factors shifts constantly. That being the case, ignoring links in the crop's production chain is ill advised.

What about RFLP probes and a DNA map for rice? Will that not speed up breeding and give farmers vastly superior rice plants? Let us hope so. But in the meantime, promises cannot feed the planet's hungry mouths. Research does not feed people, farmers do. Without extension and technology transfer, without cooperation and research networks, the world's rice equations cannot yield adequate solutions.

For Brazil, Richard Bacha and Paulo Carmona concluded that production technology and local extension had an impact on yields as great as new varieties did. In Chile, scientists considered farmer-managed research groups the key to success. So we have to be careful not to get overly attached to specific factors, whether it be biotechnology, networks, local extension, or on-farm collaboration. To be successful, extension agents need a robust set of recommendations. That presumes a good research program stands behind extension. Researchers, of course, can end up solving problems farmers do not care about, which reminds us of the importance of on-farm collaboration and applied research.

The impact the rice revolution had is not just confined to Asia's irrigated rice belt. It spread into tidal wetlands and rain-fed fields, into marginal environments in which conditions are less favorable for high yields. Even in such problematic places, much can be accomplished: by seeding rice directly in fields or by switching to varieties that mature earlier, by changing planting dates or by rotating crops. To many people who work in rice, future production depends less on biotechnology and more on closing the yield gap: between experimental fields and a farmer's paddy, between favored and marginal environments.

How to close the gap is a many-sided affair. Farmers need good seed and enough fertilizer at the right time. Crop management has to be considered—from how best to seed fields to adequate weed control. When the timing of the harvest is off, or storage facilities are inadequate, losses add up quickly. So closing the yield gap has dimensions that reach far beyond a rice field. The precise strategy, the way different components are combined, has to reflect the situation in different regions and in specific localities.

With good soil, plenty of water, and early maturing semidwarfs, farmers in the tropics can get two high-yielding rice crops. For much of monsoon Asia, however, only one rice harvest is practical. Irrigation systems depend on the runoff from watersheds, streams, and ponds, which fill up during the rainy season. Because rice needs lots of water, they can manage only one crop. But this does not preclude an alternative second crop. Mung beans have less stringent water requirements. Sometimes there is even time for a green manure like sesbania. In Pakistan's Indus Valley and across much of northern India, farmers plant rice and wheat in succession. In the savannas of South America, upland rice is part of a rotation system that includes grazing. How to control erosion, restore degraded lands, and prevent the forest's destruction are basic aspects of the problem. Rice is only one element in the equation. For monsoon Asia, by contrast, rice is the focus of agriculture—indeed, most of the time, it is agriculture.

So precisely what rice means to farmers, that is, the role rice plays in a larger production system, varies enormously from country to country. That is why national research programs are so important. No one research center can possibly be responsible for the world's rice. A collective effort is needed, but one sufficiently decentralized so that local problems get attention.

Security

I said that a book about rice does not have an ending. That is because no definitive solution exists for the kind of problem rice represents. Resistance to diseases and insect pests are often temporary attributes rather than permanent conditions. There is always another virus, blast race, or plant hopper biotype lurking. To maintain the gains already made, to keep yields from falling, requires constant vigilance. That is why research has to continue on problems presumably solved.

Fertilizer use is a good example. The impact nitrogen has on rice yields is well known. So why does IRRI bother with more research? The oil price hikes of the 1970s and early 1980s made nitrogen fertilizer, which is a petroleum-based product, much more expensive. Most Asian countries—India, Bangladesh, Burma, the Philippines—depend on imported petroleum. So what seemed solved in the days of cheap energy ended up on the problem list again. Hence the emphasis on efficiency—applying urea nitrogen at the right time, and deep enough, to nourish plants effectively. Problem solved? Not quite. Over the years, the yield response to nitrogen has dropped. Decades of rice production and heavy nitrogen use have depleted the soil. To get the same yield, farmers have to increase the nitrogen dosage. To counter this and build up the soil's structure, IRRI promotes research on nitrogen fixing plants such as azolla and sesbania.

Whether fertilizer or weeds, insects or diseases, there is no final solution. When the harvest is good, everyone gets a pat on the back. But then the cycle begins again. In agriculture, there is no end, just a pause before starting over. Success today does not mean success tomorrow. Rains can come late, frosts can come early, weeds can intrude, new pests can strike. Security rests in the collective effort needed to protect the world's rice crop. Defense rests not only on research scientists but also on local testing programs, extension teams, and farmer to farmer collaboration.

MODELS

Semidwarf varieties and new production strategies made it possible for farmers to double and triple rice yields. Cannot the strategy followed with rice be applied to other crops? Not without modifications and great caution. For rice is a special case; or more accurately, irrigated rice is a special case.

In Asia, rice is the mainstay for more than two billion people. Rice has an importance there without parallel in the Americas or Europe. To be self-sufficient in rice is a matter of state security. Given the impact modern varieties had on yields, Asian governments promoted their use. Providing rice farmers with enough seed and fertilizer became national priorities; public funds went to irrigation projects, storage facilities, national research, and local extension. Everybody had a stake in rice. Production was as much a political necessity as it was an economic priority.

In Latin America, the conversion to modern, semidwarf varieties took longer. In Brazil, for example, irrigated rice is in the country's cooler, southern states; farmers need early maturing semidwarfs with cold tolerance. In Colombia, irrigated rice is in the tropical Magdalena Valley. Although IR8 was directly introduced there, it proved susceptible to local diseases. In Latin America overall, rice has much less clout politically than in Asia. In Chile, it is a minor crop. So how fast the rice revolution reached specific countries and local farmers depended on many circumstances. For Latin America's upland rice, there is now a revolution in the making. In West Africa, the revolution has yet to come.

West Africa

Located in the Ivory Coast, the West African Rice Development Association (WARDA) was set up in 1970 to help transfer Asia's rice revolution to Africa. The attempt failed. The lesson from Asia was that success depends on semidwarf sativa rices grown with the requisite production technology. Such an apparent truth led to many false starts in West Africa.

Modern semidwarf rices are designed for irrigation. But in West Africa, taller upland types cover almost 60 percent of the cropland in rice. The remainder is mostly composed of traditional landraces—irrigated sativa types cultivated in the saturated soils of inland valleys or in brackish coastal mangrove swamps. Neither IR8 nor its successors did well in such environments. Converting inland valleys and coastal swamplands to Asian-type paddies is difficult logistically, it is very expensive, and farmers consider it undesirable. Even where irrigation is available, modern semidwarfs fall prey to local pests and diseases.[1]

Farming systems in West Africa rely much less on rice than in Asia. Root crops such as cassava and yams are much more important; so too are sorghum and millet. Moreover, farmers often intercrop. When they sow an upland rice, they intersperse rows of other crops, such as cassava and okra, with it. They do not fertilize the rice and they do not weed it.[2] Even when farmers have swampy fields suitable for modern semidwarfs, they resist the labor-intensive practices needed for high yields. That is because their labor is needed elsewhere, usually in other fields at some distance from the swamps. Under such conditions farmers do just as well with traditional rice varieties.[3]

In 1990, West Africa produced 4.7 million metric tons of rice, more than 75 percent of sub-Saharan Africa's total. Yields, however, averaged only 1.5 metric tons per hectare. Between 1970 and 1990, West African rice consumption grew by almost 6 percent a year—much faster than production. For 1988–90, annual per capita consumption of milled rice was 28 kilos; in that same period, the region imported 1.7 million tons of rice a year.[4]

Rice has great potential and a growing market in West Africa, but the Asian system of transplanted paddies is not appropriate. A combination of difficult environments, diverse cropping systems, meager inputs, and weak national programs is hard to counter. Consequently, WARDA had to start over again, designing its own breeding program from scratch. Rather than simply transfer semidwarf rices from Asia, it needed new lines adapted to African environments.

West African farmers have planted sativa rices from Asia for almost five hundred years. Most traditional African rices, especially wet rices, are now sativa cultivars. Glaberrima rices are found mostly in low-yielding upland environments. For genetic improvement, WARDA crosses traditional sativa rices from Africa with improved Asian rices. Crossing sativa and glaberrima cultivars, which are different rice species, is more difficult, as such hybrid offspring are usually sterile. Nonetheless, WARDA has made several such crosses successfully. It hopes that desirable glaberrima traits will one day diversify the genetic base of modern rice.[5]

Despite many crosses, WARDA's rice research did not have much impact until it took local farming systems seriously. Its work on mangrove-swamp environments in Sierra Leone is a case in point. The project first studied how rice fit into the local cropping system. It turned out that farmers need a range of rices with very different maturities. During the rainy season, for example, fresh water pushes out the denser, saltier marsh water for periods that range from four to six months. So the maturities required depend on how far inland a farmer's rice fields are. Moreover, the problems caused by salt-tolerant kire kire weeds are severe in the swamps as is blast disease.[6]

It took a decade to develop better alternatives. Modern semidwarfs fare poorly in the swamps, so scientists used local sativa landraces as the prototype. Such traditional rices do much better despite acid sulfate soils, high iron toxicity, and chronic saltwater intrusion. By crossing hardy traditional rices with modern ones, breeders got a high-yielding resistant rice plant that still looked like swamp rice. Using the same farming practices as before, farmers now get more than 3 metric tons of rice, up from less than 2 tons. And if they apply enough fertilizer, they can get 5 tons.[7]

The situation in West Africa reminds us that the conditions necessary for a rice revolution cannot always be duplicated. In West Africa, rice yields depend on farming systems that produce an assortment of crops. Even where rice is irrigated, as in the swamps, the degree of water salinity, blast severity, and soil types shift from one swampy inlet to the next. For rice, miracles have a political and cultural context.

Food Crops

When applied to other crops, the rice-revolution model is also misleading. Genetic improvement is not always the first step. Consider potatoes. At harvest time, farmers hold over seed tubers for the next crop. To get high yields, they need to select firm, healthy tubers and keep them as disease- and pest-free as possible. Diffused-light storage, introduced by anthropologists from the International Potato Center, can be as important to Andean potato production as better varieties. For with poor tuber selection and storage, even the best variety degenerates rapidly.[8]

Bean research at CIAT also shows that genetic improvement alone is not the key to success. The program has a surplus of improved, disease-resistant bean varieties; the problem is how to get farmers to try them. To keep costs down, farmers hold back seed at harvest time for replanting. When the same seed stock is used over many seasons, disease problems build and a cycle of low yields sets in. To promote better bean varieties and high yields, CIAT had to first tackle the seed problem. Working with cooperatives, it has helped set up community-based seed production, storage, and local distribution.[9]

With the starchy cassava root, it took CIAT a decade to make much headway. Colombia's poorest farmers produce the crop; they rarely use fertilizer or invest in crop-management improvements. Success came eventually, but because of marketing technology rather than an improved cassava plant. Plastic packaging, which doubled the grocery-store shelf life of the bulky cassava roots, was tested, and CIAT also pioneered processing cassava for animal feed. As outlets for cassava increased, prices rose and farmers showed more interest in yields.[10]

Pastures research is likewise instructive. Breeding a hardy grass that survives the dry season is just a beginning. For sustainable pastures on acidic, tropical soils, legumes must be mixed in with grasses. And the balance has to hold up despite grazing, torrential rains, and months of drought.

So rice is not a model. Nonetheless, there is an important lesson. At its best, IRRI takes an approach to local problems that is "international" yet does not infringe on national sovereignty, does not destroy initiative, and does not ignore localities. With rice, a science-based strategy is geared to the constraints poor countries and small farmers face. Using the INGER testing network, a new rice can make its way from Korea to Chile, or from Taiwan to India. Whether it is networks, training programs, or local research, whether demonstration plots or extension groups, each helps explain how a new technology can end up far from home in a form that is user-friendly to farmers.

In agriculture, one size does not fit all. At both IRRI and CIAT, the approach to rice recognizes that solutions are often local and temporary. For whether rice or potatoes, beans or cassava, there is no miraculous cultivar that can fit all the environments in which the crop is produced. Before it can be released, a new rice or potato plant has to be screened with specific diseases, insect pests, soil conditions, and abiotic stresses in mind. Nor can cultural factors, such as the timing of harvests, how crops are consumed, and farming systems, be ignored. That seems impossible. On the contrary, irrespective of the crop in question, such considerations are the building blocks of success.

What made the rice revolution possible is hard work by many people. Today, when we consider solving problems, whether it is education or acid rain, we always ask, How much? The answer is always the same: too expensive. That is because we want to do everything all at once. We have no strategy, no place to start. How much did the rice revolution cost? Between 1971 and 1984, IRRI's budget averaged $15 million a year. For the 1990s, the cost is about $30 million annually. At CIAT between 1971 and 1984, expenditures on the rice program averaged some $3 million a year. Currently, it allocates about $7 million a year to its rice program. Is that a lot? The budget for most U.S. universities tops $100 million a year. A hospital costs that much, so does a DC 10. Who paid for the rice revolution? At IRRI, the biggest donors are the United States Agency for International Development, the government of Japan, the World Bank, the Rockefeller Foundation, and the Asian Development Bank. In 1990, contributions came from more than forty countries, foundations, and international agencies. Whether it is money or people, the rice story involves us all.[11]

Success

What impact did IRRI and CIAT have on rice? Assessments reflect the definitions used. When IRRI was established, the priority was production. Asia's self-sufficiency in rice is based on varieties and technology that IRRI promoted for irrigated areas.

How much of Asia's added rice production is due to IRRI's research and how much to national programs? Apportioning success in this fashion is mostly a waste of time. After all, IRRI is supposed to cooperate with national programs, not compete with them. Why convert them into adversaries for purposes of analysis? In fact, the impact is often greatest in precisely those countries with the strongest research programs. So trying to apportion credit statistically is difficult. As to specifying the nationality of germ plasm in a modern variety, it is almost impossible. When I asked Brazilian rice breeders what percentage of the germ plasm used was "theirs," they just shook their heads. To judge from study papers published by the World Bank, "impact" is still commonly viewed in production terms: in tonnage, in higher yields, or in the number of new varieties introduced.[12] So defined, CIAT has success stories to tell about all its crop specialties. What differs is the coverage, the extent to which the success had spread. For irrigated conditions in Asia and Latin America, modern rice varieties are used almost everywhere. For sheer tonnage and impact, the rice story cannot be beat.

Once past the productivity test, new technology faced a growing list of impacts. With rice, for example, critics wanted research to stress equity. Were the benefits from new technology distributed fairly: between favored and unfavored environments, between big and small farmers, between urban consumers and rural producers, between men and women? As a definition of what impact meant, equity soon vied with production. Today, sustainable agriculture, plant genetic diversity, and integrated pest management head the list. As the items increase, so does the controversy over how to measure success.

For irrigated rice, Colombia has yields comparable to top Asian producers. In neighboring Brazil, overall yields are still below three tons. So, is CIAT's rice program a success? Divorced from the crops, countries, and environments involved, success is a slippery concept. In Colombia, where beans are a minor crop, CIAT's bean program has so far had little impact. In Costa Rica, by contrast, the Ministry of Agriculture worked with CIAT to introduce new varieties and production methods. It did so because beans are a dietary staple. As a result, Costa Rica's small farmers tried out the new varieties, liked them, and production increased rapidly.[13] So is CIAT's bean program a success? The answer is, it depends on the circumstances.

"Circumstantial success" is a good recommendation in agriculture, and good enough for the Colombian government. The budget for CIAT was cut sharply in the 1990s, as cost-conscious donor countries in the developed world reduced support. In 1994, Colombia decided to make up part of the shortfall. It contributed $1.2 million to CIAT's core budget, a figure it pledged to increase by 20 percent every year for the next five years. And circumstantial success paid off in rice. The benefits that CIAT's rice research has brought to the region is worth an estimated $600 million annually. Much of that goes to Latin America's rice consumers, who spend 40 percent less for rice today (in real terms) than in 1967. Despite such achievements, CIAT had to cut back its budget for rice research. To keep the rice program in tact, Brazil's Rice Research Agency in Rio Grande do Sul (IRGA), Colombia's FEDEARROZ, plus rice programs in Venezuela and Uruguay, set up a regional research fund in 1995. Donations of $1 million will be staggered over the next three years. Luis Roberto Sanint, current leader of CIAT's rice program, will be the fund's executive director.[14]

Clients

Although IRRI and CIAT collaborate with national programs, the available expertise is often unstable. Fortunately for IRRI, most Asian countries have a strong rice research program; nonetheless, even for Asia's rice, the situation is mixed. Research in China and Indonesia is far in advance of Burma or Vietnam. For CIAT too, country differences in research caliber are great, making collaboration difficult. In what sense can both Brazil and Bolivia be "clients" for CIAT's research? Brazil wants CIAT to concentrate on sophisticated "upstream" research, particularly in biotechnology, as it has the capacity to benefit from its subsequent diffusion. In Bolivia, by contrast, agricultural research was all but abandoned during the turbulent 1980s. So Bolivia needs CIAT for "downstream" basic training of its research staff, not for RFLP probes.

Much as rice research is organized according to irrigated, rain-fed, deepwater, and upland environments, the rice agenda at IRRI and CIAT has to consider the customer. The approach taken to networks, technology transfer, or training must be pegged to a country's research capacity and its budget—otherwise, serious mistakes follow.

BACK TO THE FUTURE

When IRRI opened in the 1960s, researchers worked without PCs, electronic mail, or anther culture. To be successful today, IRRI needs its computers and its biotechnology units. In the 1960s, testing new varieties required collaboration with national programs, with extension workers, and with local farmers. It still does.

The challenges ahead for Asia and its rice farmers are great. Prime farmland is being lost to local industries, transportation networks, and hydroelectric projects.[15] At the same time, consumer demand continues to rise. Given Asia's enormous population and its dependence on rice, it is worth reconsidering future prospects.

In the year 2030, global demand for rice is predicted to be twice that of 1990. Almost 90 percent of the additional consumption will be in Asia. Where will an extra 470 million tons of rice come from? Most is expected from irrigated paddies, the remainder from rainfed and upland environments. In South Asia, where demand will be greatest, 10 million irrigated hectares in rice-wheat rotations already show signs of fatigue. For Asia overall, the rate at which rice yields have increased has slowed: from an average of 2.8 percent a year between 1974 and 1982 to 1.8 percent for the years 1982 through 1989.[16]

According to IRRI scientists, part of the problem is that the yield ceiling for irrigated rice has not changed in thirty years. The top yield for IR8 is about 9 metric tons per hectare. No rice today can do much better than 10 metric tons. To make big gains in rice production quickly, a new rice type is needed with yields of 12 to 14 tons. Why the emphasis on irrigated rice? Because three-quarters of Asia's rice is produced in irrigated paddies. The new plant ideotype, designed by IRRI breeders in 1988–89, has fewer tillers, but the panicle on each tiller is larger, with twice as many grains. A semi-dwarf such as IR8 has twenty to twenty-five tillers, of which fifteen or so produce small panicles of one hundred grains each. The number of grains in the new type will be higher and denser; the percentage of lightweight, partially filled grains will be greatly reduced.[17]

A 1993 CGIAR rice report considered prospects for meeting consumer demand. Average yields for irrigated rice, currently about 4.5 metric tons, will need to rise to 8.5 tons. They calculate that an ultra high yielding rice could push up irrigated yields by at least 2 metric tons. The remaining 2 tons will have to come from a sharp reduction in the yield gap, from irrigating more fields, and from more double-cropping. The precise contribution each factor might make was left open. In addition, improvements across rainfed and upland environments are also necessary.[18] Can such a strategy succeed? It depends on the country in question.

Japan

Some Asian countries already have too much rice. Korea, Taiwan, and Japan all run a surplus. And their rice is expensive. Price subsidies make the crop attractive to farmers.

Japan is a case in point. I was there in 1991 to talk with Dr. Kunisuke Tanaka, an agricultural chemist at Kyoto Prefectural University. His research shows that on the average about 30 percent of the protein rice produces is not digestible.[19] The task is to find varieties with high digestible protein and get the trait into a commercial rice. He arranged for me to visit rice farms near Lake Biwa.

It was August and rice was planted everywhere, up to the edge of factories, apartment buildings, and parking lots. "Rice is part of our culture and religious traditions," Sadao Hirai told me, "part of our literature, part of our folklore." He had 3 hectares in rice, some of it special varieties for flour and rice cakes. He did not use pesticides. Transplanting took place between the end of April and the beginning of May, a ten-day period that coincided with Japan's Golden Week and included four holidays. About 80 percent of Japan's farmers are part time, and Hirai was no exception. He needed the holidays to get his rice crop in.

Japan may have enough rice, but it is running out of farmers. Hirai said that of Japan's one and a half million high school graduates, only two thousand want to be farmers. The Yamatomachi district in Niigata prefecture illustrates the problem. It is famed for the quality of its rice, but almost all its farmers are part-timers. In 1980, the district listed twenty-four hundred farmers, now it has seventeen hundred. But the number of full-time farm families is much smaller: in 1991, it was only seventy-four, and they were getting old, their average age fifty-six.

Hirai planted only half as much rice as he used to. He had diversified into potatoes, a spring crop, with melons in the summer. Nonetheless, rice was his mainstay; its subsidized price provided financial security. In 1991, the Ministry of Agriculture paid farmers $120 for a 60-kilo bag. That was ten times the world price. In the supermarket, rice sold for much less—about four times the U.S. price. Government subsidies made up the difference; in 1992, they cost Japan's taxpayers $2.5 billion.[20]

Why does Japan pay so much for rice? An end to price subsidies would undermine the profitability of part-time farming, and without part-time farming Japan would not be self-sufficient in rice. And food security is touchy political issue in Japan. The country already imports half its food, including 5 million tons of wheat from the United States. Rice is the only crop it can count on. Finally, political realities have long favored rice. Despite the population shift to the cities, Japan in 1993 had not redrawn its parliamentary districts since 1947. Compared to urban districts, rural Japan was overrepresented in Tokyo by a factor of four. So a change in Japan's rice policy has far-reaching implications.[21]

Korea and Taiwan also oppose cheap rice imports. And for some of the same reasons as Japan: for the sake of tradition, to insure food security, and because of politics.[22] In Taiwan, there simply is not another crop that can replace rice. When the monsoons hit, any lingering dry-season crops are destroyed. I watched this happen in May 1993. I was visiting the Asian Vegetable Research and Development Center (AVRDC) near Tainan. The monsoon was late, but when it hit, it struck with force. Torrential rains, terrible thunder and lightening, and heavy winds continued for almost a week. Later, I had a chance to go with Taiwanese agronomist Dr. Jocelyn Tsao to assess the damage to vegetable crops. The heavy rain had waterlogged just about everything. Much like the first frost in temperate climates, it killed all but the hardiest of plants. The next round would get the stragglers. So during the rainy season, farmers have only one choice: rice or nothing.

Rice growers in the United States want to export to Asian countries with high production costs. In 1986, the Rice Millers Association of the United States filed its first complaint against Japan. At best, however, U.S. exports can supplement, not replace, local rice production. The United States exports about 40 percent of its rice: for 1991–93, an average of 2.4 million metric tons of milled rice a year. In absolute terms, that is not very much. Although U.S. average rice production rose from 4.4 million metric tons in 1972–74 to 8.1 million in 1992–94, rice is still a minor crop domestically.[23] Rice accounts for less than 2 percent by value of U.S. crop output. Of the total harvested cropland in the United States, 24 percent goes to wheat and 25 percent to maize; rice is far behind with less than 1 percent.[24]

The United States cannot possibly substitute for Asian production. During 1992–94, Japan's rice output averaged 12.7 million tons—much more than the U.S. figure. And Japan is not a large producer by Asian standards. The average in Bangladesh is more than 27 million metric tons. So the United States cannot feed Asia; Asia will have to feed itself.[25] Can it? China, Indonesia, and India produce a combined total of more than 70 percent of Asia's rice.[26] If prospects are poor in these cases, the food situation in the next century will be grim.

China

No country produces more rice than China. And no country has a larger rural labor force: more than 780 million in 1994. Developments in rice emphasize irrigated, semidwarf varieties. Of the 33 million hectares of rice China planted in 1991, 30 million were irrigated. China also has a strong national research program. Its Ministry of Agriculture operates a system of thirty-seven agricultural research centers, most of which specialize in crop research. China developed semidwarf rice varieties before IRRI did, beginning full-scale distribution in 1964, two years before IR8 was released. By the end of the 1970s, 80 percent of China's rice land was planted to modern varieties. As elsewhere, reliance on chemical fertilizers accompanied semidwarf rices; use went up from 10 kilos per hectare in the early 1960s to 100 kilos in the early 1980s.[27]

China has maintained close ties with IRRI since the mid-1970s. Formal collaboration began in 1976 when a delegation of ten IRRI scientists spent four weeks in China. Nyle Brady, who was then IRRI's director general, led the mission. A series of reciprocal scientific missions followed. In 1978, Lin Shih-Cheng, a plant breeder at the Chinese Academy of Sciences, joined IRRI's board of trustees. In 1979, IRRI's board held its annual meeting in Beijing.[28]

The main drawback with China's semidwarfs is their susceptibility to diseases and pests. To bolster resistance, between 1971 and 1981 IRRI provided

China with more than three thousand varieties and advanced lines. Today, almost all the rice planted in China has resistance genes from IRRI stock.[29] Chinese breeding lines, in turn, have contributed valuable traits to IRRI rices, notably extra-early maturity and higher seedling vigor. Moreover, China's use of organic fertilizer, particularly azolla, has provided a promising line of research.[30]

China's economic reforms began in 1979 in agriculture. More flexible, household-based production replaced brigades and collective farming. Villages now assign land to families, and they sign contracts with local producers for poultry, pork, and fish production. Between 1978 and 1984, the amount of credit that went to agriculture tripled. The regime also promoted small, labor-intensive rural industries licensed by local town councils. Many local enterprises integrated rural production with food processing.[31] The result was greater prosperity in the countryside. Between 1979 and 1990, China's agricultural output almost doubled. Rice played a key role in that success.[32]

Almost all of China's rice, some 90 percent, is produced south of the Qinling Mountains.[33] To the north of this divide, the climate gets cooler and rainfall tapers off; soils are loose and calcareous. The north is China's wheat belt. South of the Qinlings, which includes the Yangtze Valley, China gets hotter and wetter. Soils are sticky, which makes them ideal for rice.[34] During 1979–80, China produced an average of 144 million metric tons of rice; during 1989–90, it produced more than 185 million.[35] The annual growth rate in rice production averaged 2.8 percent a year, compared to 1.5 percent for the country's population.[36] China increased double-cropping, used more nitrogen fertilizer, promoted green manure, and intensified mixed farming, particularly fish production in rice paddies.[37] It also managed to push up the yield ceiling for irrigated rice. Assigning land to families was the key social factor behind China's success. Its development of hybrid rice seed was the key technical innovation.

Rice is an inbred, self-fertilizing crop. Each plant has both male and female organs. To cross two varieties, breeders remove one plant's male parts, fertilizing its remaining female parts with pollen from the other. This is practical for breeding purposes, but not to produce millions of tons of seed. Hybrid vigor, however, can increase yields by 15 to 20 percent. So it is a very desirable trait.[38] Chinese scientists found a way to mass-produce hybrid seed. In 1964, they discovered a wild *Oryza* species with natural male sterility. By incorporating the male sterility trait into parent rices used in hybrid crosses, it became unnecessary to emasculate thousands of plants. Between

1981 and 1991, the use of hybrid seed on rice acreage in China increased from 6 to over 15 million hectares.[39] China and IRRI now collaborate on hybrid seed technology. They continue to exchange rice germ plasm and have joint breeding projects on disease resistance, pest tolerance, and yield potential.[40]

Indonesia

In the 1970s, Robert Chandler analyzed what made a rice production program successful. He concluded that new varieties and production technology are not enough. Countries need a good credit system, an extension program, storage facilities, and farmer-led organizations.[41] In 1985, IRRI celebrated its twenty-fifth anniversary. Just about every country in Asia sent a delegation. According to A. Affandi, Indonesia's Minister of Agriculture, that lesson had not changed.[42]

Rice is Indonesia's main crop. In 1990, rice was grown on 10.5 million hectares—three times the acreage that went to maize, which was second in importance, and eight times the acreage devoted to soybeans. More than 100 million people, some 20 million rural households, depend on rice for their livelihood. They are small landholders, with rarely more than 2 hectares and often less than 1.[43]

In 1964, before IRRI released IR8, Indonesia harvested 12.3 million tons of rice, not much more than the 11.7 million tons it had produced a decade earlier. Between 1970 and 1975, it imported a total of 5.6 million metric tons of rice. A decade later, the situation had greatly improved. Indonesia was self-sufficient in rice; it grew 40 million tons—more than three times the 1964 figure. Yields had increased from 1.8 to 4 tons.[44] What had so improved the country's track record?

Public-sector investment is part of the story. Between 1969 and 1985, Indonesia built or rehabilitated 2.8 million hectares of irrigated land. For 1975 through 1979, the government allocated 23 percent of its development budget to agriculture.[45] According to Affandi, however, the key to success was how Indonesia's Rice Intensification Program worked with farmers. The project grouped three or more villages into units. Each unit had field extension workers assigned to it, plus a village unit bank for credit. Village unit cooperatives distributed seed, fertilizer, and equipment; they also handled postharvest processing and marketing. The project promoted modern varieties, careful water management, plant protection, and better soil preparation.[46] An oil-rich country, Indonesia heavily subsidized the cost of fertilizer. Nitrogen use on rice fields rose from an average of 34 kilos of urea per hectare in 1971 to 183 kilos in 1986.[47]

The area included in the intensification program increased from 1.6 million hectares in 1968 to 8.6 million in 1984. By then, there were more than 200,000 village groups involved in extension. Farmers with small plots grouped together so they could work the land more efficiently. When brown plant hoppers got out of control, more resistant varieties from IRRI were quickly adopted. Strong, local organization made it possible to replace insecticides with monitored pest management.[48]

Indonesia also has close ties to IRRI. Henry Beachell, after he retired as head of IRRI's breeding program in 1972, took an outreach assignment there, a post he held until 1981. Since 1970, IRRI has trained more than five hundred Indonesian researchers.[49]

India

Both China and Indonesia profited from high-yielding modern rices. So did India, but not as much. India has more acreage in rice than any country in the world, but it is not the largest producer. In 1994, India harvested 118 million metric tons of rice on 42 million hectares. China had much less acreage in rice, approximately 30 million hectares, but it produced 178 million tons—over 40 percent more.[50] The difference is due mostly to rice environments and production technology.

India irrigates only 45 percent of its rice acreage, compared to 72 percent for Indonesia and 93 percent for China. Much of India's rice land, almost 40 percent, is in either a rain-fed or a flood-prone environment. Breeding high-yielding rices for these production systems is difficult. In 1991, only 65 percent of India's rice acreage was planted to modern varieties, compared to 85 percent for Indonesia and virtually 100 percent in China's case. India has been self-sufficient in rice since 1977, but overall yields are still only 2.6 metric tons, far below China's average of 5.7 tons or Indonesia's 4.4 tons.[51] Given such constraints, how can India's rice production keep pace with demand?

In states such as Punjab in the north and Tamil Nadu in the south, farmers irrigate almost all their rice; yields are about 5 metric tons, which is comparable to Indonesia's.[52] Nonetheless, gains can still be made from irrigated rice. According to Indian officials, the average yield from 105 irrigated rice districts across 14 states was just 3 metric tons. With better management, yields could increase at least a ton, adding 15 million metric tons to India's rice output.[53] Nonetheless, to produce enough rice for the future, more will have to come from rainfed areas like Eastern India. Is that possible?

India has the research capacity and extension experience it needs for success. In 1993, it had 28 agricultural universities, 48 research institutes, 30 national research centers, and 80 all-India crop improvement programs, which involved both research and technology transfer.[54] India also maintains the most extensive rice germ-plasm collection in Asia; it is a center of biodiversity for many crops, including indica rice.[55]

Agriculture is a top priority in India. In 1994, the sector still provided direct employment for two-thirds of the working population; it accounted for a third of India's gross domestic product and for a quarter of its exports.[56]

In eastern India, monsoon rainfall averages more than 2,000 millimeters. The region accounts for 58 percent of India's rice acreage, only 20 percent of which is irrigated, and currently produces about 45 percent of India's rice. To keep the country self-sufficient, India has targeted the region's rain-fed environments, setting up some fifteen hundred special production districts. The emphasis is on "location-specific technology" developed through farmer participation and on-farm research. This includes converting from low-yielding monsoon production to winter rice. In 1990, "frontline" demonstration projects were in operation on 10,000 hectares. According to officials, both improved rain-fed rices and new production technology for harsh environments are available. What India needs is a "mission-oriented" effort to transfer them speedily and effectively.[57]

The production program for eastern India is multifaceted. It involves input supply and rice marketing, farmer education, dependable credit, facilities for sharing farm equipment, and local seed production—all of which depend on a committed extension service. Long-term improvements on the land include better drainage, grading and consolidation, rainwater collection in upland areas, along with tanks and wells for lowland irrigation.[58]

Given its diverse cultures, farming systems, and rice environments, India's self-sufficiency in rice is a great accomplishment. Its location-specific approach is a key ingredient in that success. For what seems an absolute at the office can be highly relative in the field.

In 1994, I spent several weeks at the International Crops Research Institute for the Semi-Arid Tropics (ICRISAT) in the South Indian state of Andhra Pradesh. During field trips, I had a chance to visit rice paddies. The terrain and physical layout was not what I expected. In the Philippines, Luzon's central valley has rice fields everywhere. In Japan and Taiwan, my experience was likewise based on contiguous paddies. South of Hyderabad, by contrast, the land was rocky, with cliffs and rugged hills. Farmers had to fit in small paddies between piles of crags and boulders, wherever they found level land or could clear enough space. Trees were everywhere too, not just on the rocky ground between paddies but often right smack in the middle of them.

At Santa Barbara in the Philippines, as in most of Japan, farmers puddle fields and transplant rice within a couple hectic weeks. In much of Andra Pradesh, however, "access to irrigation" has a relative rather than an absolute dimension. At project sites, about 20 percent of the paddies were already transplanted, about half were still covered with weeds, and the rest were being harrowed or puddled.

Compared to the Philippines, South China, or Taiwan, the monsoon rains in Andhra Pradesh are not so dependable or as heavy. So farmers take their time. It was early July and they were unsure just how strong the rains would be. We stopped to talk with a farmer and his son. They were preparing a paddy with the help of their water buffalo and were knee deep in mud.[59] For irrigation, they had an uncovered pit well about 14 meters deep. The farmer explained that if the monsoon held up, he would plant 3 hectares, all the land he had, in rice. The rain plus his well water would be enough to sustain it. If the rains tapered off, however, he would plant sorghum in half his fields—a crop he does not have to irrigate. He estimated that if he planted 3 hectares of rice and the monsoon stopped abruptly, he would loose 1 hectare, as he had only enough well water for 2 hectares of rice.

Can Asia feed itself? is the kind of great question global thinkers like to ponder. It is easy to forget the answer depends on the local calculus farmers use to make decisions about millions of rice paddies all over Asia. Whether Asia can feed itself depends on national production plans, on local research programs, and on extension—just as it did when IRRI started. It depends on how strong the research chain is that links together the work at IRRI with national programs. It depends on commitment, whether of farmers in Andhra Pradesh or molecular biologists at IRRI. In the case of Asia's biggest rice-producing countries, they know the stake they have in rice and in the countryside. They have the local organization in millions of villages that makes revolution possible.

DEVELOPMENTS

Asia had a record-breaking rice harvest in 1990: an increase of 7.5 percent, or 34 million metric tons, over 1988. For China, the increase came to 11.5

percent; for Indonesia it was 8.4 percent; and for India, 5.2 percent. Since then, the pace at which Asia adds to its rice stocks has leveled. Compared to 1990, when Asia harvested 480 million metric tons of rice, the three-year average for 1992–94 was only 482 million tons, hardly any increase at all. For Indonesia, harvests for 1992–94 were up by an average of 2.4 million metric tons over 1990, or by 5.3 percent; for India, the average increase came to 2.9 million tons, or 2.5 percent. In China, however, the amount of rice harvested declined by an average of 9 million tons, or by 4.7 percent. For 1995, China's rice production increased an estimated 11 million tons over 1994, but the total was still below the record harvest of 1990.[60]

In 1994, National Public Radio reported on a "super rice" in Asia. It may have arrived just in time. According to the press, IRRI prototypes are expected to yield from 12 to 12.5 metric tons per hectare, an increase of 20 to 25 percent over the best existing varieties. The new rices type have only 6 to 8 tillers but all are productive and each has a large panicle with 200 to 250 grains. To support the heavy panicles, they have thicker, sturdier stems with dark green erect leaves. They have a vigorous root system and mature in just 100 to 130 days. The new rice type is especially good for direct seeding, which will save on labor and water. Its genetic background differs from most modern rices. Breeders started out by crossing tropical java rices from Indonesia, a group closer genetically to japonica rices than to indica rices. Hybrids will be developed by IRRI by crossing the new plant type with modern high-yielding varieties. Hybrid vigor could raise yields to more than 13 metric tons per hectare.[61]

To develop the new rice, IRRI has made more than one thousand crosses since 1990, generating some fifty thousand breeding lines. According to Dr. Gurdev Khush, the project's director, much work lies ahead. A quiet, determined man, Khush headed the IRRI breeding team that developed IR36. When I wrote to him about "super rices" in 1995, he was cautious. He expects the yield advantage to hold up in trials at different sites in Asia. Further improvements are needed in disease and pest resistance; then national programs will have to select for taste, size, and cooking quality. The IRRI project is not alone. At Indonesia's Research Institute for Rice, Dr. Zainuddin Harahap has developed high-yielding javonica lines as promising as IRRI's. A breakthrough in yields would make the future of Asia's rice and its farmers more secure.[62]

The motto at IRRI for the 1990s is to "do more with less." And not only because the institute's budget was cut. Asia is losing land to industries and urban growth; more rice will have to come from less land. For a sustainable future, Asia's farmers will need rices that can be grown with less water, less fertilizer, and less pesticides. The research agenda at IRRI reflects the new realities.[63]

More with less seems an appropriate policy for a billion Asian rice farmers to follow. In developed countries, urban consumers do the opposite; they use as much as possible and get back less than ever.

Sustainable Development

Who is responsible for sustainable development? All of us. It is not just a job for IRRI and CIAT. The waste and overconsumption of rich countries threaten the planet's collective future as much as the demand for food in poor countries. In the end, the planet's development is as much about "us" as it is about "them."

Every year, estimates chemist Paul Connett, the United States throws away 1.6 billion pens, 2 billion disposable razors, 16 billion paper diapers, and 22 billion plastic grocery bags.[64] Considering all types of packaging, from plastic boxes for pins and tacks to special thermal wrapping for fast food, Americans throw out some 270 kilos (600 pounds) of it per person, which sums to a staggering 76 billion kilos (168 billion pounds) annually.[65] For Americans, sustainable development begins at home. Reuse and recycling has to be applied in neighborhoods, in communities, at schools, and at work.

The rice revolution started on tiny plots of land all over Asia. To be successful, farmers had to change customary production technology. For the sake of our common future, it is time for us urban farmers to do our part.

NOTES

INTRODUCTION

1. United Nations, *Demographic Yearbook 1991* (New York: United Nations, 1992), p. 103.

2. For statistics on Asia's population growth and rice production, see H. M. Beachell, "The Need for a Global Rice Research System," in *Progress in Irrigated Rice Research*, by International Rice Research Institute (IRRI) (Manila: IRRI, 1989), pp. 315–324.

3. Norman E. Borlaug, "Challenges for Global Food and Fiber Production," *K. Shogs-o. Lantbr.akad. Tidskr.* Supp. 21 (1988): 20.

4. See Cynthia Hewitt de Alcantara, *Modernizing Mexican Agriculture: Socioeconomic Implications of Technological Change 1940–1970* (Geneva: United Nations Research Institute for Social Development, 1976), pp. 19–36. Also see Everett M. Rodgers, "Evolution and Transfer of the U.S. Extension Model," in *The Transformation of International Agricultural Research and Development*, ed. Lin Compton (Boulder, Colo.: Lynne Rienner, 1989), pp. 137–52.

5. See D. S. Athwal, "Semidwarf Rice and Wheat in Global Food Needs," *Quarterly Review of Biology* 46 (March 1971): 8–11, and Dana G. Dalrymple, *Development and Spread of High-Yielding Wheat Varieties in Developing Countries* (Washington, D.C.: Agency for International Development, 1986), pp. 13–19. For data on Mexico's wheat yields and production, see Food and Agricultural Organization (FAO) of the United Nations, *FAO Production Yearbook 1960* (Rome: FAO, 1961), p. 35 and FAO, *Food Production Yearbook 1972* (Rome: FAO, 1973), p. 52.

6. For the 1960 data, see FAO, *Food Production Yearbook 1960* (Rome: FAO, 1961), p. 35; for 1970, see FAO, *Food Production Yearbook 1972* (Rome: FAO, 1973), p. 52; for 1987, see FAO, *Food Production Yearbook 1989* (Rome: FAO, 1990), p. 116; and for 1988–90, see FAO, *Food Production Yearbook 1990* (Rome: FAO, 1991), p. 70. Also see Borlaug, "Challenges," pp. 24–25; generally, see pp. 23–29.

7. Borlaug, "Challenges," pp. 35, 42; also see Caroll P. Streeter, *A Partnership to Improve Food Production in India* (New York: Rockefeller Foundation, 1969), pp. 8–24.

8. On how serious this seemed geopolitically, see John Keery King, "Rice Politics," *Foreign Affairs* (April 1953): 453–60.

9. Robert F. Chandler Jr., *An Adventure in Applied Science: A History of the International Rice Research Insitute* (1982; reprint, Manila: IRRI, 1992), pp. 4–5; generally, see pp. 1–9.

10. Ibid., pp. 20–25, and Appendix Six, p. 208. For a critical view of IRRI's early years, see Robert S. Anderson, Edwin Levy, and Barrie M. Morrison, *Rice Science and Development Politics: Research Strategies and IRRI's Technologies Confront Asian Diversity 1950–1980* (Oxford: Clarendon Press, 1991), pp. 22–110.

11. Chandler, *Adventure*, pp. 155–56 and Hewitt de Alcantara, *Mexican Agriculture*, pp. 46–47. The acronym CIMMYT stands for Centro Internacional de Mejoramiento de Maíz y Trigo.

12. Norman E. Borlaug and Christopher R. Dowswell, "World Revolution in Agriculture," *Encyclopedia Britannica Yearbook*, 1985, p. 8.

13. Borlaug, "Challenges," p. 42.

14. IRRI, *World Rice Statistics 1987* (Manila: IRRI, 1988), p. 8.

15. Chandler, *Adventure*, p. 156.

16. The acronym CIAT stands for Centro Internacional de Agricultura Tropical. For a discussion of crops originating in the Americas, see Jack R. Harlan, *Crops and Man*, 2d. ed. (Madison, Wisc.: American Society of Agronomy, 1992), pp. 217–36.

17. Chandler, *Adventure*, p. 159.

18. The story behind the Centers and the Consultative Group is told by Warren C. Baum, *Partners Against Hunger* (Washington, D.C.: World Bank, 1986). Also see Chandler, *Adventure*, pp. 159–65.

19. Consultative Group on International Agricultural Research (CGIAR), *CGIAR Annual Report 1984* (Washington, D.C.: CGIAR Secretariat, 1985), p. 21.

20. See James Lang, *Inside Development in Latin America: A Report from the Dominican Republic, Colombia, and Brazil* (Chapel Hill: University of North Carolina Press, 1988).

21. The centers publish books, research articles, and conference reports on the crops they study. Much needed material came from Winrock International's Agribookstore in Arlington, Virginia. Unfortunately, the bookstore's services were discontinued in 1995.

CHAPTER ONE

1. International Rice Research Institute (IRRI), *IRRI Rice Almanac, 1993–1995* (Manila: IRRI, 1993), pp. 7, 127. The population figure is for 1994; see U.S. Bureau of the Census, *World Population Profile: 1994* (Washington, D.C.: Government Printing Office, 1994), p. A-4.

2. The statistics on wheat, rice, and maize acreage, and the 1992–94 rice production figures, are compiled from Food and Agricultural Organization (FAO) of the United Nations, *FAO Production Yearbook 1994* (Rome: FAO, 1995), pp. 68–69, 70–71, and 77–78. Unless otherwise indicated, all figures are for "rough" rice rather than "milled" rice. Milling removes the hull and bran from threshed, rough rice. The U.S. conversion rate of rough to milled rice is .70; for a list of conversion rates, see IRRI, *World Rice Statistics 1990* (Manila: IRRI, 1991), p. 315.

3. The percentage of total rice production exported is compiled from tables in FAO, *FAO Trade Yearbook 1993* (Rome: FAO, 1994), p. 95, and *FAO Production Yearbook 1993* (Rome: FAO, 1994), p. 70. For an overview of basic rice facts, see Cristina C. David, "The World Rice Economy: Challenges Ahead," in G. S. Khush and G. H. Toenniessen, eds., *Rice Biotechnology* (Wallingford, England: CAB International, 1993), pp. 1–18. On cereal exports, see Robert E. Huke and Eleanor H. Huke, *Rice: Then and Now* (Manila: IRRI, 1990), p. 33, and Thomas R. Hargrove, "A Grass Called Rice," mimeograph (Los Baños: IRRI, 1991), pp. 2–3. For human consumption of cereals, see IRRI, *Rice Almanac*, p. 7.

4. Per capita U.S. rice consumption in 1993 was 17.6 pounds, excluding quantities used in alcoholic beverages; see U.S. Department of Agriculture (USDA), *Agricultural Statistics 1994* (Washington, D.C.: Government Printing Office, 1994), p. 41. When the rice used in beer and processed food is added in, the average is ten kilos; see IRRI, *Rice Almanac*, p. 103. Caloric intake for 1990 is from IRRI, *Rice Almanac*, p. 129; the figure cited includes Bangladesh, Myanmar (Burma), Cambodia, Laos, and Vietnam.

5. Huke and Huke, *Rice*, pp. 18, 13, and David, "World Rice Economy," p. 1. For farm size, also see IRRI, *World Rice Statistics 1990*, p. 220.

6. See Joseph K. Campbell, *Dibble Sticks, Donkeys, and Diesels: Machines in Crop Production* (Manila: IRRI, 1990), pp. 21–22.

7. The description of comb harrows follows Campbell, *Dibble Sticks*, p. 124.

8. Rice production practices are many and varied; for an overview of rice environments and the many factors that shape production, see Duane S. Mikkelsen and Surajit K. De Datta, "Rice Cultures," in *Rice*, ed. Bor S. Luh, 2 vols. (New York: Van Norstrand Reinhold, 1991), 1:103–86; D. H. Grist, *Rice* (Whitstable: Longman, 1975), pp. 139–56; and Surajit K. De Datta, *Principles and Practices of Rice Production* (New York: John Wiley & Sons, 1981), pp. 221–55. For cultural practices prior to the green revolution, consult V. D. Wickizer and M. K. Bennett, *The Rice Economy of Monsoon Asia* (Stanford, Calif.: Stanford University Press, 1941), pp. 31–62. For China, Korea, and Japan at the turn of the century, see F. K. King, *Farmers of Forty Centuries* (Madison, Wisc.: Democratic Printing, 1911).

9. Based on data cited in Huke and Huke, *Rice*, pp. 28–29.

10. Ibid., p. 13.

11. This section is based on interviews with Te-Tzu Chang, IRRI, 8 August 1990, pp. 120–23 and Duncan Vaughan, IRRI, 8 August 1990, pp. 117–19. All interviews are available at Vanderbilt University Library. They are organized by center, by the national program involved, and by local agencies. Pages are numbered sequentially. Also see Te-Tzu Chang, "The Rice Cultures," *Philosophical Transactions of the Royal Society of London* B. 275, (1976): 143–57, and Te-Tzu Chang, "Domestication and Spread of Cultivated Rices," in *Foraging and Farming—The Evolution of Plant Exploitation*, ed. D. R. Harris and G. C. Hillman (London: Unwin Hyman, 1989), pp. 408–17.

12. Escavations in Zhejiang Province in the Yangtze Delta found rice grains that have been dated to 7000 B.P.; see Hiko-Ichi Oka, "Genetic Diversity," in Khush and Toenniessen, *Rice Biotechnology*, p. 74. Recent evidence suggests an even earlier date; see Peter T. White, "Rice the Essential Harvest," *National Geographic* (May 1994): 65–66.

13. On submergence and floating rice, see Grist, *Rice*, pp. 42–43 and 140–41; also see Mikkelsen and De Datta, "Rice Cultures," pp. 160–62. For wet rice and dry rice types, see Francesca Bray, *The Rice Economies: Technology and Development in Asian Societies* (Oxford: Basil Blackwell, 1986), p. 11.

14.On the origins of rice and its diversity, see Bray, *Rice Economies*, pp. 8–19, and Lucien M. Hanks, *Rice and Man* (Chicago: Aldine Atherton, 1972), pp. 16–22. For the effects of solar radiation, see Mikkelsen and De Datta, "Rice Cultures," pp. 122–23, and De Datta, *Rice Production*, pp. 22–25.

15. For a technical discussion of indica-japonica divergence, see Hiko-Ichi Oka, "The Origin of Cultivated Rice and Its Adaptive Evolution," in *Rice in Asia*, by Association of Japanese Agricultural Scientific Societies (Tokyo: University of Tokyo Press, 1975), pp. 21–32.

16. For characteristics of indica and japonica rices, see Grist, *Rice*, pp. 93–98. Grist notes that Indonesian varieties are intermediate and sometimes classified separately as *javonica* rices. On regional preferences, see Dana G. Dalrymple, *Development and Spread of High-Yielding Rice Varieties in Developing Countries* (Washington, D.C.:

Agency for International Development, 1986), p. 4, and White, "Essential Harvest," p. 63.

17. On photoperiod response, see Mikkelsen and De Datta, "Rice Cultures," pp. 121–22; Grist, *Rice*, pp. 116–17; and De Datta, *Rice Production*, p. 25.

18. This paragraph is based on Hiko-Ichi Oka, "Genetic Diversity of Wild and Cultivated Rice," in Khush and Toenniessen, *Rice Biotechnology*, p. 66. The description of seeding practices in various rice environments also draws on De Datta, *Rice Production*, pp. 235, 238, 241, 247–48.

19. See M. S. Swaminathan, "Rice," *Scientific American* 250 (1984): 84; generally, see pp. 81–85, and Hargrove, "Grass Called Rice," p. 2.

20. IRRI, *Annual Report 1992–1993: Rice in Crucial Environments* (Los Baños: IRRI, 1993), p. 21.

21. On the origin of African rices, consult Jack R. Harlan, *Crops and Man*, 2d ed. (Madison, Wisc.: American Society of Agronomy, 1992), pp. 92, 138, 182, 189–90. On rice in West Africa, see David C. Littlefield, *Rice and Slaves: Ethnicity and the Slave Trade in Colonial South Carolina* (Baton Rouge: Louisiana State University Press, 1981), pp. 80–98.

22. For the Middle Ages, see Maguelonne Toussaint-Samat, *A History of Food* (Cambridge: Blackwell, 1992), pp. 161–62; generally, see pp. 153–63. For Lombardy, see Fernand Braudel, *The Mediterranean and the Mediterranean World in the Age of Philip II*, 2 vols. (New York: Harper & Row, 1975), 1:69, 74. For Italy's current production, see IRRI, *World Rice Statistics 1990*, p. 6. Italy's preference for japonica rices is noted in Dalrymple, *Development and Spread*, pp. 15–16.

23. Dauril Alden, *Royal Government in Colonial Brazil* (Berkeley and Los Angeles: University of California Press, 1968), pp. 364–66, and IRRI, *World Rice Statistics 1990*, p. 4. An arroba varied in weight between twenty-five to thirty-two pounds; see Alden, *Royal Government*, p. xxii.

24. Braudel, *Mediterranean* 1:326; Magnus Morner, "Rural Economy and Society in Spanish America," in *Colonial Spanish America*, ed. Leslie Bethell (Cambridge, Eng.: Cambridge University Press, 1987), p. 303; and IRRI, *Rice Almanac*, p. 130.

25. Littlefield, *Rice and Slaves*, pp. 95–98, 108–14, and Herman J. Viola and Carolyn Margolis, eds., *Seeds of Change* (Washington, D.C.: Smithsonian Institution Press, 1991), pp. 165–66. For Georgia, see Julia Floyd Smith, *Slavery and Rice Culture in Low Country Georgia, 1750–1860* (Knoxville: University of Tennessee Press, 1985).

26. For a summary of rice cultivation statistics for the years 1839 to 1889, see Department of the Interior, Census Office, *Report on the Statistics of Agriculture in the United States at the Eleventh Census: 1890* (Washington, D.C.: Government Printing Office, 1895), pp. 71–72. For 1993 production figures by state, see U.S. Department of Agriculture, *Agricultural Statistics 1993* (Washington, D.C.: Government Printing Office, 1993), p. 20.

27. Hargrove, "Grass Called Rice," p. 2, IRRI, *Rice Almanac*, p. 31, and FAO, *Production Yearbook 1994*, p. 71.

28. IRRI, *Rice Almanac*, p. 31.

29. On U.S. rice production, see IRRI, *Rice Almanac*, pp. 103–4. For the size of rice farms in the United States and Japan, see Patrick Smith, "Letter from Tokyo," *New Yorker*, 14 October 1991, p. 108.

30. For statistics on how rice is consumed, see *Rice Almanac*, pp. 103, 105, 129. On rice in beer, also see White, "Essential Harvest," p. 54, and Tom Hargrove, *A Dragon Lives Forever: War and Rice in Vietnam's Mekong Delta, 1969–1991 and Beyond* (New York: Ivy Books, 1994), p. 344.

31. FAO, *Production Yearbook 1994*, p. 70.

32. Compiled from tables in United Nations, *Demographic Yearbook 1991* (New York: United Nations, 1992), p. 103.

33. How serious the crisis was can be seen by reviewing William Paddock and Paul Paddock, *Famine—1975!* (Boston: Little, Brown, 1967), and George Borgstrom, *The Hungry Planet* (New York: Macmillan, 1965).

34. Production figures are compiled from IRRI, *Rice Statistics 1990*, pp. 2–3; for population figures, see United Nations, *Demographic Yearbook 1966* (New York: United Nations, 1967), p. 124.

35. IRRI, *Rice Statistics 1990*, pp. 2–3. On rice production in China, also see Frank Leeming, *The Changing Geography of China* (Oxford: Blackwell, 1993), pp. 73–94.

36. See Robert F. Chandler Jr., *An Adventure in Applied Science: A History of the International Rice Research Institute* (1982; reprint, Manila: IRRI, 1992), pp. 51–55. For the contrast between rice types and the background to IR8, see Robert F. Chandler Jr., *Rice in the Tropics: A Guide to the Development of National Programs* (Boulder, Colo.: Westview, 1979), pp. 12–13, 32–35.

37. The quote is from Chandler, *Adventure*, pp. 53–54; on the development of IR8, also see pp. 103–11. The story behind the semidwarf rice varieties is ably told by D. S. Athwal, "Semidwarf Rice and Wheat in Global Food Needs," *Quarterly Review of Biology* 46 (March 1971): 1–30. Also see Dalrymple, *Development and Spread*, pp. 15–19, and Randolph Barker, Robert Herdt, with Beth Rose, *The Rice Economy of Asia* (Washington D.C.: Resources for the Future, 1985), pp. 54–72.

38. On China, see Dalrymple, *Development and Spread*, p. 18 and De Datta, *Rice Production*, pp. 179–80.

39. On the advantages and disadvantages of IR8, see Chandler, *Rice in the Tropics*, pp. 35–37, and Chandler, *Adventure*, p. 109.

40. On IR20, see Hargrove, *Dragon Lives Forever*, p. 110.

41. On IR36, see Hargrove, "Grass Called Rice," p. 6. For a case study of IR36, see Donald L. Plucknett, Nigel J. H. Smith, J. T. Williams, and N. Murthi Anishetty, *Gene Banks and the World's Food Supply* (Princeton, N.J.: Princeton University Press, 1987), pp. 171–84.

42. Hargrove, "Grass Called Rice," p. 6. For an excellent summary of breeding efforts at IRRI, see De Datta, *Rice Production*, pp. 173–220. Also see Gurdev S. Khush, "Rice Breeding: Past, Present, Future," *Journal of Genetics* 66 (1987): 195–216.

43. Chandler, *Adventure*, pp. 111–12.

44. Figures on irrigated acreage are compiled from *Rice Statistics 1990*, p. 212. Asia's irrigated area and production is compiled from tables in IRRI, *Rice Almanac*, pp. 133, 135. The story of how modern rice varieties spread is told in Dalrymple, *High-Yielding Rice Varieties*, and in Barker, Herdt, and Rose, *Rice Economy*, pp. 54–72.

45. IRRI, *Rice Statistics 1990*, pp. 14, 8.

46. For the statistics on irrigation and the use of modern varieties, see IRRI, *Rice*

Almanac, pp. 61–63, 46–48, 58–60, and 55–57. Averages for 1992–94 rice production are compiled from tables in FAO, *Production Yearbook 1994*, pp. 70–71.

Japan's yields are normally above 6 metric tons. Unseasonably cold weather made 1993 a disasterous year; yields were just 4.6 metric tons, which pulls down the 1992–94 average.

47. IRRI, *Rice Statistics 1990*, p. 14, and FAO, *Production Yearbook 1994*, p. 70.

48. Barker, Herdt, and Rose, *Rice Economy*, p. 77.

49. IRRI, *Rice Almanac*, p. 112.

50. Barker, Herdt, and Rose, *Rice Economy*, p. 62, generally pp. 62–64.

51. Based on Chandler, *Rice in the Tropics*, pp. 103–9.

52. IRRI, *World Rice Statistics 1990*, p. 2.

53. Chandler, *Rice in the Tropics*, pp. 109–13. Also see T. H. Shen, *Agricultural Development on Taiwan Since World War II* (Ithaca, N.Y.: Cornell University Press, 1964), pp. 36–54, 88–103, 340–74.

54. Chandler, *Rice in the Tropics*, pp. 113–18; for South Korea's rice imports, see IRRI, *World Rice Statistics 1990*, p.35.

55. Based on Chandler, *Rice in the Tropics*, pp. 119–23. Subsequent assessments were not so positive. Burmeister felt the diffusion campaign for Tongil rice adoption was forced on farmers, despite their opposition and the reservations of Korean rice scientists. In 1980, a combination of cold weather and blast suceptibility led to a 30 percent drop in the harvest. Thereafter, the extension service promoted a Tongil-japonica mix; see Larry L. Burmeister, *Research, Realpolitic, and Development in Korea: The State and the Green Revolution* (Boulder, Colo.: Westview Press, 1988), pp. 50–73.

56. Figures on rice prices and per capita consumption are from IRRI, *Crucial Environments*, p. 4.

57. The preceding conclusions are based on Michael Lipton and Richard Longhurst, *Modern Varieties, International Agricultural Research, and the Poor* (Washington, D.C.: World Bank, 1985), pp. 23–47. For a recent assessment, see the introduction in Cristina C. David and Keijiro Otsuka, eds., *Modern Rice Technology and Income Distribution in Asia* (Boulder, Colo.: Lynne Rienner, 1994), pp. 3–21.

58. See Peter B. R. Hazell and C. Ramasamy, *The Green Revolution Reconsidered: The Impact of High-Yielding Rice Varieties in South India* (Baltimore, Md.: Johns Hopkins University Press, 1991), pp. 240–45, 4–5.

59. See Rita Sharma and Thomas T. Poleman, *The New Economies of India's Rice Revolution: Income and Employment Diffusion in Uttar Pradesh* (Ithaca, N.Y.: Cornell University Press, 1993), pp. 9–10, 14–16, and generally pp. 1–22, 239–55.

60. See Cristina C. David and Keijiro Otsuka, "Modern Rice Technology: Emerging Views and Policy Implications," in David and Otsuka, *Modern Rice Technology*, pp. 421–22, 448, and generally pp. 411–29. On regional disparities in India, also see C. Ramasamy, P. Paramasivam, and A. Kandaswamy, "Irrigation Quality, Modern Variety Adoption, and Income Distribution: The Case of Tamil Nadu in India," in David and Otsuka, *Modern Rice Technology*, pp. 323–73, and Sharma and Poleman, *New Economics*, pp. 1–21.

61. For population figures, see United Nations, *Demographic Yearbook 1992* (New York: United Nations, 1993), p. 146.

62. For a recent discussion of population growth, see Amartya Sen, "Population: Dilusion and Reality," *New York Review* (22 September 1994): 62–70.

63. For population growth rates in the 1960s, see U.S. Bureau of the Census, *Statistical Abstract of the United States 1971* (Washington, D.C.: Government Printing Office, 1971), pp. 794–96, and *Statistical Abstract of the United States 1991*, pp. 830–32. For population estimates, see IRRI, *IRRI: Towards 2000 and Beyond* (Manila: IRRI, 1989), p. 4.

64. For Asia's 1994 rice production, see FAO, *Production Yearbook 1994*, p. 70. The projections are outlined in IRRI, *Towards 2000*, pp. 1–5, and Hargrove, "Grass Called Rice," pp. 9–10, 12. The estimate for 2020 presumes that Asia will still account for about 90 percent of the world's rice production.

65. For an excellent summary of varietal improvement, rice ecologies, and production problems, see De Datta, *Rice Production*, pp. 173–545. The classic text on rice breeding is P. R. Jennings, W. R. Coffman, and H. E. Kauffman, *Rice Improvement* (Los Baños: IRRI, 1979).

66. On how important rice quality is in Asia, see L. J. Unnevehr, B. Duff, and B. O. Juliano, eds., *Consumer Demand for Grain Quality* (Ottawa: International Development Research Council, 1992).

67. See Chandler, *Adventure*, pp. 176–77.

68. For a brief review of IRRI's testing network, see IRRI, *International Rice Research: 25 Years of Partnership* (Los Baños: IRRI, 1985), pp. 66–68.

69. What follows is based on an interview with D. C. Seshu, IRRI, 24 July 1990, pp. 18–21. Seshu has since left IRRI for the International Center for Living Aquatic Resources Management in Manila, Philippines. Also see D. V. Seshu, "Agricultural Research in Networks: A Model for Success," in *Vegetable Research in Southeast Asia*, by Asian Vegetable Research and Development Center (AVRDC) (Taipei: AVRDC, 1988), pp. 211–18 and D. V. Seshu, "Varietal Performance in International Irrigated Rice Trials," in *Rice Farming Systems: New Directions*, by IRRI (Manila: IRRI, 1989): 113–27.

70. See the schedule for planting times in Mikkelsen and De Datta, "Rice Cultures," pp. 110–18.

71. See IRRI, *Crucial Environments*, p. 41, and Michael Jackson and Robert Huggan, "Sharing the Diversity of Rice to Feed the World," *Diversity* 9, no. 3 (1993): 25.

72. IRRI, *Annual Report 1991–1992: Sharing Responsibilities* (Manila: IRRI, 1992), p. 54.

73. Based on IRRI, *Rice Almanac*, p. 119.

74. What follows is based on an interview with Susan McCouch, IRRI, 26 July 1990, pp. 35–38. Responsibility for any errors or oversimplification is mine, certainly not Dr. McCouch's. For a discussion of RFLP research, see Susan R. McCouch and S. D. Tanksley, "Development and the Use of Restriction Length Polymorphism in Rice Breeding and Genetics," in Khush and Toenniessen, *Rice Biotechnology*, pp. 109–33, and the articles in IRRI, *Rice Genetics II* (Manila: IRRI, 1991), pp. 435–501. McCouch currently collaborates with IRRI as a shuttle scientist. For a recent review of IRRI's work in biotechnology, see IRRI, *Annual Report 1991–1992*, pp. 13–14, and IRRI, *Rice Almanac*, pp. 118–23.

75. Based on IRRI, *Rice Almanac*, p. 120.

76. On selecting parents for crosses, see S. D. Tanksley, N. Ahn, M. Causse, R.

Coffman, T. Fulton, S. R. McCough, G. Second, T. Tai, Z. Wang, K. Wu, and Z. Yu, "RFLP Mapping of the Rice Genome," in IRRI, *Rice Genetics II*, pp. 435–38; on tagging projects and IRRI's coordinating role, see Gary H. Toenniessen and Gurdev S. Khush, "Prospects for the Future," in Khush and Toenniessen, *Rice Biotechnology*, p. 311.

77. IRRI, *Rice Almanac*, p. 120.

78. See W. Ronnie Coffman, Johnson Olufowote, Pablo Grau, and Reynaldo Villareal, "A New Approach to Rice Improvement in Latin America," in *Rice in Latin America: Improvement, Management, and Marketing*, ed. Frederico Cuevas-Pérez (Cali: CIAT, 1992), p. 37.

79. What follows is based on an interview with D. S. Brar, IRRI, 25 July 1990, pp. 21–24.

80. What follows is based on an interview with D. Senadhira, IRRI, 27 July 1990, pp. 46–48. On salt tolerance, also see IRRI, *Highlights 1987* (Los Baños: IRRI, 1988), pp. 18–20.

81. Based on IRRI, *Rice Almanac*, pp. 118–19.

82. What follows is based on an interview with D. Senadhira, IRRI, 13 August 1990. On varietal improvement at IRRI, including cold tolerance, see IRRI, *International Rice Research*, pp. 49–74.

83. See IRRI, *Towards 2000*, p. 26.

84. The information below is based on an interview with HilleRisLambers, IRRI, 10 August 1990, pp. 123–28. Also see R. Thakur and D. HilleRisLambers, "Nondestructive Techniques for Measuring Plant and Internode Elongation," in *Proceedings of the 1987 International Deepwater Workshop*, by IRRI (Manila: IRRI, 1988), pp. 209–14. HilleRisLambers left IRRI in 1993.

85. See IRRI, *Rice Almanac*, p. 133.

86. Based on an interview with Michel Arraudeau, IRRI, 11 August 1990, pp. 92–95. For statistics on Indonesia's upland rice production, see IRRI, *Rice Almanac*, p. 60. For basic facts about upland rice production, see M. A. Arraudeau and B. S. Vergara, *A Farmer's Primer on Growing Upland Rice* (Manila: IRRI, 1988).

87. What follows is based on an interview with Te-Tzu Chang, IRRI, 15 August 1990, pp. 129–34. Chang retired from IRRI in 1991 and was succeeded by Dr. Michael Jackson of England. Also see Te-Tzu Chang, "Conservation of Rice Genetic Resources: Luxury or Necessity?" *Science* 224 (20 April 1984): 251–56, and Te-Tzu Chang, "Principles of Genetic Conservation," *Iowa State Journal of Research* 59 (May 1985): 325–48. In general, see Plucknett et al., *Gene Banks*.

88. IRRI, *Annual Report 1993–1994: Filling the World's Rice Bowl* (Manila: IRRI, 1994), p. 37.

89. See the list in D. A. Vaughan, "Passport Data on Rice and Related Species," in *Rice Germplasm, Collecting, Preservation, Use*, by IRRI (Manila: IRRI, 1991), pp. 59–65.

90. IRRI, *Annual Report 1993–1994*, p. 37.

91. Chandler, *Adventure*, pp. 103–4, and Jackson and Huggan, "Sharing Diversity," pp. 22–25. For IRRI's past work on collection, see IRRI, *Rice Germplasm*, pp. 5–6. The Consultative Group on International Agricultural Research (CGIAR), which represents IRRI, has endorsed the International Convention on Biological Diversity. According to recent agreements, the germ-plasm collection at IRRI and other centers is held in trusteeship under the auspices of the Commission on Plant Genetic Resources of the Food and

Agricultural Organization (FAO) of the United Nations; see "World News," *Diversity* 9, nos. 1–2 (1993): 4–6.

92. Jackson and Huggan, "Sharing Diversity," p. 23. The problem of genetic erosion is not confined to rice; it effects most cereal crops as well as vegetable and fruit species; see Cary Fowler and Pat Mooney, *Shattering: Food, Politics, and the Loss of Genetic Diversity* (Tucson: University of Arizona Press, 1990).

93. Based on an interview with Duncan Vaughan, IRRI, 8 August 1990, pp. 117–19. Vaughan left IRRI in 1993 to join the genetic resources unit of Japan's national agricultural research center at Tskuba.

94. Jackson and Huggan, "Sharing Diversity," p. 23.

95. See Vaughan, "Passport Data," p. 61.

96. See Hargrove, *Dragon Lives Forever*, pp. 302–3.

97. See IRRI, *Rice Almanac*, pp. 114–18.

98. This does not include deepwater rice or tidal wetlands. For a classification of rain-fed rice lands and the acreage they comprise, see IRRI, *Terminology for Rice Growing Environments* (Los Baños: IRRI, 1984), pp. 7–8. Also see the breakdown in IRRI, *Towards 2000*, p. 26.

99. Based on an interview with David Mackill, IRRI, 26 July 1990, pp. 40–44, and with S. K. De Datta, IRRI, 8 August 1990, pp. 107–9. Mackill left IRRI in 1991, De Datta in 1992. Also see S. K. De Datta, "Technology Development and the Spread of Direct-Seeded Flooded Rice in Southeast Asia," *Expl Agric.* 22 (1986): 417–26, and W. R. Coffman and J. S. Nanda, "Cultivar Development for Dry Seeded Rice," in *Cropping Systems Research in Asia*, by IRRI (Los Baños: IRRI, 1982), pp. 149–56. Also, see the articles in IRRI, *Direct-Seeded Flooded Rice in the Tropics* (Manila: IRRI, 1991).

100. For the contrast between dry- and wet-seeded rice, see IRRI, *Annual Report 1993–1994*, p. 23, and Mikkelsen and De Datta, "Rice Cultures," pp. 134–70.

101. On direct seeding and weed problems, see IRRI, *Annual Report 1993–1994*, pp. 24, 26. For statistics on U.S. herbicide use on rice, see USDA, *Agricultural Statistics 1993*, pp. 464–65.

102. Based on an interview with Keith Moody, IRRI, 15 August 1990, pp. 156–61. Also see Keith Moody, S. K. De Datta, V. M. Bhan, and G. B. Manna, "Weed Control in Rainfed Lowland Rice," in *Progress in Rainfed Lowland Rice*, by IRRI (Los Baños: IRRI, 1986), pp. 359–70, and S. K. De Datta, P. C. Bernasor, T. R. Migo, M. A. Llagas, and P. Nantasomsaran, "Emerging Weed Control Technology for Broadcast Seeded Rice," in *Progress in Irrigated Rice Research*, by IRRI (Los Baños: IRRI, 1989), pp. 133–46; K. Moody, "Weed Control in Dry Seeded Rice," in IRRI, *Cropping Systems*, pp. 161–78; and Kwesi Ampong-Nyarko and S. K. De Datta, *A Handbook for Weed Control in Rice* (Manila: IRRI, 1991). Also consult D. E. Bayer, "Weed Management," in Luh, *Rice* 1:287–307.

103. Based on an interview with Pablo Escuro, PhilRice, IRRI, 10 August 1990, pp. 176–78. For an overview of weed control problems, including dry-seeded rice, see De Datta, *Rice Production*, pp. 460–512.

104. IRRI, *Rice Statistics 1990*, p. 15.

105. Based on an interview with Graeme Quick, IRRI, 24 July 1990, pp. 9–13. Quick left

IRRI in 1994. Also see IRRI, *International Rice Research*, pp. 99–104. For a discussion of postharvest technology, see De Datta, *Rice Production*, pp. 513–45.

106. For illustrations and descriptions of threshing techniques and equipment, see Campbell, *Dibble Sticks*, pp. 199–205.

107. Chandler, *Rice in the Tropics*, p. 65.

108. The description that follows is based on Chandler, *Rice in the Tropics*, pp. 65–90. Also see James I. Wadsworth, "Milling," in Luh, *Rice* 1:379–81.

109. For the nutritional value of rice, see Chandler, *Rice in the Tropics*, pp. 9–12. On the merits of polished versus brown rice, see Hargrove, *Dragon Lives Forever*, pp. 343–44, and Chandler, *Rice in the Tropics*, pp. 81, 85. For a description of modern rice drying techniques and storage technology see C. Y. Wang and Bor S. Luh, "Harvest, Drying, and Storage of Rough Rice," in Luh, *Rice* 2:311–45.

110. Agrilectric Power Company, "Converting Rice Husks to Energy," pamphlet (1988), pp. 1–6. Also see White, "Essential Harvest," p. 63.

111. See Hargrove, "Grass Called Rice," p. 7. Most of what follows is based on an interview with Ronald Buresh, IRRI, 9 August 1990, pp. 69–101. Buresh returned to the United States in 1991. An overview of fertilizer problems can be found in De Datta, *Rice Production*, pp. 348–419. Also see IRRI, *International Rice Research*, pp. 82–90. On nitrogen fixation by the rice plant itself, see G. S. Khush and J. Bennett, eds., *Nodulation and Nitrogen Fixation* (Manila: IRRI, 1993).

112. For oil prices, see *Statistical Abstract of the United States: 1991*, pp. 754, 698.

113. Based on interviews with Jagdish Ladha, IRRI, 25 July 1990, pp. 25–31, and W. Ventura, IRRI, 25 July 1990, pp. 32–34. Also see J. K. Ladha and I. Watanabe, "Biochemical Basis of 'Azolla-Anabaena azolla' Symbiosis," in *Azolla Utilization*, by IRRI (Los Baños: IRRI, 1987), pp. 47–58. Also see the azolla articles in S. K. Dutta and Charles Sloger, *Biological Nitrogen Fixation Associated with Rice Production* (Washington, D.C.: Howard University Press, 1991), pp. 63–118.

114. See Zhang Zhuang-ta, Ke Yu-si, Ling De-quan, Duan Bing-yuan, and Liu Xi-lian, "Utilization of 'Azolla' in Agricultural Production in Guangdong Province, China," in IRRI, *Azolla Utilization*, pp. 141–45.

115. For recent articles, see J. K. Ladha, T. George, and B. B. Bohlool, eds., *Biological Nitrogen Fixation for Sustainable Agriculture* (Boston: Kluwer Academic Publishers, 1992), and Dutta and Sloger, *Biological Nitrogen Fixation*, pp. 279–366.

116. See J. K. Ladha, I. Watanabe, and S. Saono, "Nitrogen Fixation by Leguminous Green Manure and Practices for Its Enhancement in Tropical Lowland Rice," in *Green Manure in Rice Farming*, by IRRI (Los Baños: IRRI, 1988), pp. 165–83. Also see R. K. Pandey, *A Primer on Organic-Based Rice Farming* (Manila: IRRI, 1991).

117. Based on an interview with Kong Hoeng, IRRI, 27 July 1990, pp. 55–59. Also see IRRI, *Brown Planthopper: Threat to Rice Production in Asia* (Los Baños: IRRI, 1979). For an overview of insect problems, see M. O. Way and C. C. Bowling, "Insect Pests of Rice," in Luh, *Rice* 1:237–67.

118. For Indonesia's experience with plant hoppers, see A. Affandi, "The Rice Revolution in Indonesia: The Indonesian Experience in Increasing Rice Production," in *Impact of Science on Rice*, by IRRI (Manila: IRRI, 1985), pp. 34–37, and Christopher Vaughan, "Disarming Farming's Chemical Warriors," *Science News* 134 (20 August 1988): 120–21. For an

excellent analysis of the role farmer field schools played, see Elske van de Fliert, *Integrated Pest Management: Farmer Field Schools Generate Sustainable Practices: A Case Study in Central Java Evaluating IPM Training* (Wageningen, Netherlands: Wageningen Agricultural University, 1993). On IR36 and plant-hopper resistance, see Plucknett et al., *Gene Banks*, pp. 173–79. On the pesticide-induced resurgence of plant hoppers, see Agnes C. Rola and Prabhu L. Pingali, *Pesticides, Rice Productivity, and Farmers' Health: An Economic Assessment* (Manila: IRRI, 1993), pp. 15–18, 20–21. On pesticide resistance in insects, see Fowler and Mooney, *Shattering*, pp. 48–49. Pest control using natural enemies is by no means an IRRI innovation; see the many examples in Paul Debach and David Rosen, *Biological Control of Natural Enemies*, 2d ed. (Cambridge, England: At the University Press, 1991).

119. In general, see P. S. Teng and K. L. Hoeng, eds., *Pesticide Management and Integrated Pest Management in Southeast Asia* (College Park, Md.: Consortium for International Crop Protection, 1989), and W. H. Reissig, E. A. Heinrichs, J. A. Litsinger, K. Moody, L. Fiedler, T. W. Mew, and A. T. Barrion, *Illustrated Guide to Integrated Pest Management in Rice in Tropical Asia* (Los Baños: IRRI, 1986).

120. See Rola and Pingali, *Pesticides*, pp. 19–21, 7–9, and 1–3; some of the figures the authors cite are drawn from Barker, Herdt, and Rose, *Rice Economy*, p. 89. For the United States, see USDA, *Agricultural Statistics 1993*, pp. 464–65.

121. Rola and Pingali, *Pesticides*, pp. 31–40. The health data is from IRRI, *Annual Report 1993–1994*, pp. 22–23. In general, see G. Forget, T. Goodman, and A. De Villiers, *Impact of Pesticide Use on Health in Developing Countries* (Ottawa: International Development Research Center, 1993).

122. Based on an interview with Ramesh Saxena, IRRI, 3 August 1990, pp. 102–6. Saxena left IRRI in 1991. Also see R. C. Saxena, "Insecticides from Neem," in *Insecticides of Plant Origin*, ed. J. T. Arnason, B. J. R. Philogene, and P. Morand (Washington, D.C.: ASC Symposium Series, 1989), pp. 110–35.

123. The uses of neem are based on National Research Council, *Neem: A Tree for Solving Global Problems* (Washington, D.C.: National Academy Press, 1992), pp. 12–13, 42–43, 60–77. Although neem has great potential, it does kill nontarget insects; see L. G. Soon and D. G. Bottrell, *Neem Pesticides in Rice* (Manila: IRRI, 1994). On the difficulties of protecting rice in storage, see Robert R. Cogburn, "Insect Pests of Stored Rice," in Luh, *Rice* 1:269–85.

124. Based on an interview with Paul Teng, IRRI, 13 August 1990, pp. 147–51. For an overview of disease problems, see T. W. Mew, "Rice Diseases," in Luh, *Rice* 1:187–236.

125. See Dalrymple, *High-Yielding Rice Varieties*, pp. 42–44, and IRRI, *Rice Almanac*, pp. 48, 130.

126. See the report on sheath blight in IRRI, *Annual Report 1993–1994*, p. 20.

127. Ibid.

128. Based on an interview with Sadiqul I. Bhuiyan, IRRI, 17 August 1990, pp. 194–98. Also see S. M. Miranda, "Irrigation System Principles and Practices for Reliable and Efficient Water Supply to Rice Farms," in IRRI, *Progress in Irrigated Rice*, pp. 193–202. For a review of water-management problems in rice, see De Datta, *Rice Production*, pp. 297–347.

129. For a practical guide, see Hugues Dupriez and Phillippe deLeener, *Ways of Water: Run-Off, Irrigation, and Drainage* (London: Macmillan, 1992).

130. Based on an interview with Keith Ingram, IRRI, 9 August 1990, pp. 162–67. Ingram left IRRI in 1993.

131. IRRI, *Annual Report 1993–1994*, pp. 83–84.

132. What follows is based on an interview with Pal Singh, IRRI, 30 July 1990, pp. 87–91. Also see V. P. Singh et al., eds., *Training Resources Book for Farming Systems Diagnosis* (Los Baños: IRRI, 1990). For a discussion of on-farm research and its impact, see M. Zahidul Hoque, *Cropping Systems in Asia: On Farm Research and Management* (Los Baños: IRRI, 1984).

133. In general, see J. L. Brewbaker and N. Glover, "Woody Species as Green Manure Crops in Rice-Based Cropping Systems," in IRRI, *Green Manure*, pp. 29–43.

134. Based on an interview with Sam Fujisaka, IRRI, 14 August 1990, pp. 152–55. Fujisaka left IRRI for CIAT in 1994. Also see Sam Fujisaka, "A Method For Farmer Participatory Research and Technology Transfer: Upland Soil Conservation in the Philippines," *Expl Agric.* 25 (1989): 423–33, and S. Fujisaka, "The Need to Build Upon Farmer Practice and Knowledge: Reminders from Selected Upland Conservation Projects and Policies," *Agroforestry Systems* 9 (1989): 141–53. In general, also see Joyce Moock and Robert Rhoades, eds., *Diversity, Farmer Knowledge, and Sustainability* (Ithaca, N.Y.: Cornell, 1992).

135. Based on interviews and notes compiled at Santa Barbara; see IRRI fieldnotes, 31 July 1990–1 August 1990, pp. 64–78.

136. Based on notes compiled at Guimba and on an interview with Hermenegildo Gines, IRRI, 1 August 1990, pp. 79–85. Also see Dennis P. Garrity and Hermenegildo C. Gines, "The Development of Rice-Corn Rotations in Tropical Lowland Environments: A Systems Research Approach" (paper presented at the International Seminar on Integrated Technology of Field Corn Production in Paddy Fields, 26–31 March 1990, University of the Philippines at Los Baños, Laguna, Philippines).

137. See Hargrove, "Grass Called Rice," p. 8; IRRI, *Crucial Environments*, pp. 41–43; IRRI, *Rice Almanac*, p. 109; and IRRI, *Annual Report 1991–1992*, pp. 10–11.

138. IRRI, *Crucial Environments*, p. 45.

139. Benito S. Vergara, *A Farmer's Primer on Growing Rice* (Los Baños: IRRI, 1979); Hargrove, "Grass Called Rice," p. 8.

140. IRRI, *Annual Report 1991–1992*, pp. 4, 9.

141. IRRI, *Annual Report 1993–1994*, p. 45.

142. Ibid.

143. See Hargrove, *Dragon Lives Forever*, pp. 335–36, and IRRI, *World Rice Statistics 1990*, p. 8.

144. IRRI, *Annual Report 1991–1992*, pp. 18, 59.

145. Based on an interview with Klaus Lampe, IRRI, 17 August 1990, pp. 191–93. In April 1995, George S. Rothschild, an Australian entomologist, became IRRI's new director general.

CHAPTER TWO

1. For example, see Emiko Obnuki-Tierney, *Rice as Self: Japanese Identities through Time* (Princeton, N.J.: Princeton University Press, 1993).

2. For per capita rice consumption in selected Latin American countries, see International Rice Research Institute (IRRI), *IRRI Rice Almanac 1993–1995* (Manila: IRRI, 1993), p. 129. Most modern rices in Latin America are based on indica crosses; see Dana Dalrymple, *Development and Spread of High-Yielding Rice Varieties in Developing Countries* (Washington, D.C.: Agency for International Development, 1986), pp. 79–90. Also see the grouping of parents for CIAT-ICA crosses in the informational bulletin of CIAT's rice program, *Arroz en las Américas* (April 1992): 7. In the 1930s, there were japonica rices in Rio Grande do Sul; they were not much liked by consumers, who considered them too sticky; see Paulo Carmona, "Evolución del cultivo de arroz con riego en Rio Grande do Sul, Brasil," *Arroz en las Américas* (December 1984): 2.

3. For per capita rice consumption in Brazil, see IRRI, *Rice Almanac*, p. 129. The U.S. figure is for 1992; see U.S. Department of Agriculture (USDA), *Agricultural Statistics 1993* (Washington D.C.: Government Printing Office, 1993), p. 46.

4. On farm size in Asia, see IRRI, *World Rice Statistics 1990* (Manila: IRRI, 1991), p. 220. For the definition of "small" used by Colombia's rural development agency, see James Lang, *Inside Development in Latin America: A Report from the Dominican Republic, Colombia, and Brazil* (Chapel Hill: University of North Carolina Press, 1988), pp. 101, 105–7.

5. For an excellent overview of the region's agrarian policies, see Merilee S. Grindle, *State and Countryside: Development Policy and Agrarian Politics in Latin America* (Baltimore, Md.: Johns Hopkins University Press, 1986). See also Gerson Gomes and Antonio Pérez, "The Process of Modernization in Latin American Agriculture," *CEPAL Review* 8 (1979): 55–74.

6. Compiled from the rates in Luis Lopéz Cordovez, "Trends and Recent Changes in the Latin American Food and Agricultural Situation," *CEPAL Review* 16 (1982): p. 14.

7. Joint Agricultural Division, "The Agricultural of Latin America: Changes, Trends, and Outlines of Strategies," *CEPAL Review* 27 (1985): 121.

8. Inter-American Development Bank (IDB), *Economic and Social Progress in Latin America, 1986 Report* (Washington, D.C.: IDB, 1986), p. 125.

9. For the percentage of Brazil's population categorized as urban and rural, see Fundação Instituto Brasileiro de Geografia e Estatística (IBGE), *Anuário Estatístico do Brasil 1982* (Rio de Janeiro: IBGE, 1983), p. 116; for metropolitan São Paulo, see ibid., p. 86. Cities near the million mark in 1970 and 1980 are calculated from the municipal population of each census year rather than from the population of metropolitan regions; see ibid., pp. 86–87. Subsequent citations of the *Anuário* are by year.

10. Andrew Hacker, ed., *U/S: A Statistical Portrait of the American People* (New York: Viking Press, 1983), p. 24.

11. IBGE, *Anuário Estatístico do Brasil 1980*, pp. 73, 77.

12. The exception is Nicaragua, whose population was 38 percent rural in 1993; see World Bank, *World Development Report 1995* (New York: Oxford University Press, 1995), pp. 222–23.

13. López Cordovez, "Trends and Recent Changes," p. 20.

14. Ibid., pp. 37–38. See also Manuel Figueroa L., "Rural Development and Urban Programming," *CEPAL Review* 25 (1985): 111–27.

15. López Cordovez, "Trends and Recent Changes," p. 20.

16. For the 1965 urbanization figures, see World Bank, *World Development Report 1990* (New York: Oxford University Press, 1990), p. 238; for 1980, see *World Development Report 1982* (New York: Oxford University Press, 1982), p. 148; for 1993, see *World Development Report 1995* (New York: Oxford University Press, 1995), p. 222.

17. López Cordovez, "Trends and Recent Changes," p. 26. For Brazil, see Gary Howe and David Goodman, *Smallholders and Structural Change in the Brazilian Economy: Opportunities in Rural Poverty Alleviation* (San José: International Fund for Agricultural Development, 1992), pp. 33–55. On beans and maize, the staple crops of Mexico's small farmers, see Steven E. Sanderson, *The Transformation of Mexico's Agriculture: International Structure and Politics of Rural Change* (Princeton, N.J.: Princeton University Press, 1986), pp. 214–29.

18. World Bank, *World Development Report 1991*, pp. 278–79.

19. Food and Agricultural Organization of the United Nations (FAO), FAO *Production Yearbook 1994* (Rome: FAO, 1995), p. 3.

20. Compiled from FAO, *Production Yearbook 1994*, p. 70; the figure subtracts out U.S. production but includes the Caribbean.

21. IRRI, *World Rice Statistics 1990*, pp. 4, 16.

22. For Brazil's population, see IBGE, *Anuário Estatístico do Brasil 1971*, p. 39 and IBGE, *Anuário Estatístico do Brasil 1991*, p. 180.

23. For total rice acreage by type of environment, see IRRI, *World Rice Statistics 1990*, p. 217 and IRRI, *Rice Almanac*, p. 133.

24. Compiled from IBGE, *Anuário Estatístico do Brasil 1993*, pp. 3–28, 3–35, and IBGE, *Anuário Estatístico do Brasil 1994*, pp. 3–27, 3–35.

25. On immigration to Brasil, see Thomas W. Merrick and Douglas H. Graham, *Population and Economic Development in Brazil, 1800 to the Present* (Baltimore, Md.: Johns Hopkins University Press, 1979), pp. 80–117.

26. What follows is based on interviews and field trips with Richard Bacha, EMPACE, 13–14 February 1989, pp. 2–14. In general, also see Milton Geraldo Ramos, ed., *Manual de Produção de Arroz Irrigado* (Florianópolis: EMPASC, 1985).

27. The acronym EMPASC stands for Empresa Catarinense de Pesquisa Agropecuária.

28. On parboiling in rice, see Bor S. Luh and Robert R. Mickus, "Parboiled Rice," in Bor S. Luh, ed., *Rice*, 2 vols. (New York: Van Nostrand Reinhold, 1991), 2:51–88, and Surajit K. De Datta, *Principles and Practices of Rice Production* (New York: John Wiley & Sons, 1981), pp. 526–29.

29. IBGE, *Anuário Estatístico do Brasil 1976*, p. 168, *Anuário Estatístico do Brasil 1985*, p. 343, *Anuário Estatístico do Brasil 1986*, p. 279, and *Anuário Estatístico do Brasil 1994*, pp. 3–28, 3–35.

30. Empresa Catarinense de Pesquisa Agropecuária (EMPASC), *Recomendação de Cultivares para o Estado de Santa Catarina* (Florianópolis: EMPASC, 1988), p. 16.

31. IRRI, *Rice Almanac*, p. 40.

32. Compiled from *Anuário Estatístico do Brasil 1993*, pp. 3–28, 3–35, and IBGE, *Anuário Estatístico do Brasil 1994*, pp. 3–28, 3–35.

33. Instituto Rio Grandense do Arroz (IRGA), *Anuário Estatístico de Arroz 1986* (Porto Alegre: IRGA, 1988), pp. 11–13, 109.

34. Ibid., p. 12–13.

35. Ibid., pp. 16–18.

36. Ibid., p. 10.

37. The acronym IRGA stands for Instituto Rio Grandense do Arroz.

38. See CIAT, *Arroz en las Américas* 13 (April 1992): 2.

39. What follows is based on interviews with Paulo Carmona, IRGA, 15 February 1989, pp. 23–24.

40. Paulo Carmona, "Evolución de arroz," pp. 1–4.

41. Compiled from IRGA, *Anuário Estatístico de Arroz 1986*, p. 107.

42. On bluebelle, see Dalrymple, *Development and Spread*, p. 79.

43. IRGA, *Anuário Estatístico do Arroz 1986*, p. 107.

44. Ibid., pp. 34, 109.

45. Based on an interview with Maurício Fisher, IRGA, 15 February 1989, pp. 34–37.

46. IRRI, *Rice Almanac*, p. 40.

47. On the classification of red rice and hybridization with commerical varieties, see Fabio A. Montealegre and José Patricio Vargas, "Management and Characterization of Red Rice in Colombia," in *Rice in Latin America: Improvement, Management, and Marketing*, ed. Frederico Cuevas-Pérez (Cali: CIAT, 1992), pp. 119, 128–35.

48. Also see Montealegre and Vargas, "Management of Red Rice," pp. 120–27. Red rice is a problem in the United States, and many control methods are similar; see Roy J. Smith, "Integrated Red Rice Management," in Cuevas-Pérez, *Rice in Latin America*, pp. 143–58.

49. What follows is based on field-trip notes and interviews with Paulo Carmona, IRGA, 16–17 February 1989, pp. 38–45.

50. Based on interviews with Fernando Bruno, IRGA, 23 February 1989, pp. 46–49, and with Fernandinho Dominguez, CPATB, 22 February 1989, pp. 11–12.

51. The acronym CPATB stands for Centro de Pesquisa Agropecuária de Terras Baixas.

52. See IRRI, *Rice Almanac*, p. 40. The state's total rice area can vary considerably from year to year. In 1990, for example, farmers harvested 698,000 hectares of rice, and in 1991, 804,000 hectares; see *Anuário Estatístico do Brasil 1991*, p. 517, and *Anuário Estatístico do Brasil 1992*, p. 553.

53. What follows is based on an interview with José Carlos Reis, CTATB, 22 February 1989, pp. 4–6.

54. See R. K. Pandey, *A Farmer's Primer on Growing Soybean on Riceland* (Manila: IRRI, 1987), pp. 61–65. For the 1990–93 soybean data, see FAO, *Production Yearbook 1993*, p. 106.

55. What follows is based on an interview with Francisco de Jesus Vernetti, CTATB, 22 February 1989, pp. 7–10 and Vanderlie da Rosa Caetano, CTATB, 22 February 1989, pp. 9–10.

56. Based on an interview with Angelo Soares, IRGA, 24 February 1989, pp. 50–54.

57. The acronym CPAC stands for Centro de Pesquisa Agropecuária do Cerrado.

58. What follows is based on an interview with Carlos Magno Campos da Rocha, CPAC, 9 August 1988, pp. 1–5.

59. Compiled from figures for soybean production in Mato Grosso, Goiás, Mato Grosso de Sul, and the Federal District; see IBGE, *Anuário Estatístico de Brasil 1987–88*, p. 341. For cattle, see IBGE, *Anuário Estatístico do Brasil 1977*, pp. 161, 165. The 1990 figures are from IBGE, *Anuário Estatístico do Brasil 1991*, pp. 542, 546.

60. Compiled from IBGE, *Anuário Estatístico do Brasil 1971*, p. 149.

61. For articles on Brazil's research system, see Levon Yeganiantz, ed., *Brazilian Agriculture and Agricultural Research* (Brasilia: Department of Diffusion of Technology, 1984).

62. The acronym CNPAF stands for Centro Nacional de Pesquisa de Arroz e Feijão.

63. This estimate subtracts out acreage in Rio Grande do Sul and Santa Catarina from Brazil's total. It also assumes that no more than 20 percent of the acreage remaining was irrigated; see IBGE, *Anuário Estatístico do Brasil 1987–88*, pp. 332–333, 343, and IBGE, *Anuário Estatístico do Brasil 1991*, p. 510. Similar information can be found in IRRI, *Rice Almanac*, pp. 42, 133.

64. For a breakdown of irrigated and upland rice production for 1985 though 1989, see Guia Rural Magazine, "Lugar de arroz no brejo," *Guia Rural* (December 1989): 48.

65. I have added 80 percent of the rice production in Minas Gerais to the cerrado total; see IBGE, *Anuário Estatístico do Brasil 1987–88*, pp. 332–333, 343, and IBGE, *Anuário Estatístico do Brasil 1991*, p. 510.

66. IRGA, *Anuário Estatístico do Arroz 1986*, p. 109, IBGE, *Anuário Estatístico do Brasil 1971*, p. 149, and IBGE, *Anuário Estatístico do Brasil 1986*, p. 279.

67. Compiled from IBGE, *Anuário Estatístico do Brasil 1992*, p. 546, and IBGE, *Anuário Estatístico do Brasil 1994*, pp. 3–28 and 3–35.

68. Based on an interview with Reinaldo de Paula Ferreira, CNPAF, 14 March 1989, pp. 9–10.

69. See the interview with Sônia Milagres Teixeira, CNPAF, 17 March 1989, pp. 19–20.

70. This and many subsequent observations on Brazil's upland rice production are also noted in IRRI, *Rice Almanac*, pp. 42–43.

71. Based on an interview with Luis Fernando Stone, CNPAF, 13 March 1989, pp. 15–18.

72. The importance of deep plowing and a dense root system is also noted in M. S. Arraudeau and B. S. Vergara, *A Farmer's Primer on Growing Upland Rice* (Manila: IRRI, 1988), pp. 54–70.

73. See C. H. Kingsolver, T. H. Barksdale, and M. A. Marchetti, *Rice Blast Epidimiology* Bulletin 853 (September 1984), Pennsylvania State University College of Agriculture, Agricultural Experimental Station, University Park, Pennsylvania, pp. 1–25, and De Datta, *Rice Production*, p. 40.

74. See the interview with A. S. Prabhu, CNPAF, 16 March 1989, pp. 11–12.

75. Based on an interview with Evane Ferreira, CNPAF, 16 March 1989, pp. 15–18. Also see Evane Ferreira and José Francisco da Silva Martins, *Insectos Prejudiciais ao Arroz no Brasil e Seu Controle* (Goiânia: CNPAF, 1984), pp. 10–11, 27–28, 39–41.

76. What follows is based on an interview with Emílio da Maia Castro, CNPAF, 14 March 1989, pp. 2–8.

77. During the 1980s, research budgets declined in many Latin American countries; see W. Ronnie Coffman, Elicio P. Guimarães, and César P. Martinez, "New Approaches to Rice Improvement," in Cuevas-Pérez, *Rice in Latin America*, pp. 34–35.

78. The acronym EMGOPA stands for Empresa Goiás de Pesquisa Agrícola. What follows is based on the author's interview with Waldemar Pinto Cerqueira, CNPAF, 15 March 1989, pp. 20–23.

CHAPTER THREE

1. For Chile's 1994 population, see U.S. Bureau of the Census, *World Population Profile: 1994* (Washington, D.C.: Government Printing Office, 1994), p. A-6.

2. For rainfall and temperature data, see James W. Wilkie and Adam Perkal, eds., *Statistical Abstract of Latin America* (SALA) (Los Angeles: UCLA Latin American Center Publications, 1984), 23:62.

3. The classic statement is M. F. Millikan and D. Hopgood, *No Easy Harvest: The Dilemma of Agriculture in Underdeveloped Countries* (Boston: Little, Brown, 1967; see especially pp. x, 19–21, and 67–77. Also see Thedore W. Schults, *Transforming Traditional Agriculture* (New Haven, Conn.: Yale University Press, 1964), pp. 175–206, and Grace E. Goodell, "Bugs, Bunds, Banks, and Bottlenecks: Organizational Contradictions in the New Rice Technology," *Economic Development and Cultural Change* 33 (October 1984): 23–41. A balanced approach is also emphasized in Arthur T. Mosher, *Getting Agriculture Moving* (New York: Praeger, 1966).

4. The acronym INIA stands for Instituto Nacional de Investigaciones Agropecuarias. For Chile's 1992–1994 rice production, see Food and Agricultural Organization (FAO) of the United Nations, *FAO Production Yearbook 1994* (Rome: FAO, 1995), p. 70.

5. What follows is based on an interview with Luis Becerra, INIA, 15 March 1990, pp. 7–10. On the organization of Technology Transfer Groups (GTT), see Luis Becerra R., *El ABC de los G.T.T.* (Quilamapu: INIA, 1988).

6. *SALA* 23:309; *SALA* 27:553, 550.

7. *SALA* 29:780–81, 890, and *SALA* 27:553.

8. Based on an interview with Carlos Lago, INIA, 15 March 1990, pp. 11–14.

9. What follows is based on an interview with Mario Mellado, INIA, 16 March 1990, pp. 22–24.

10. Figures on per capita wheat consumption are from Centro Internacional de Mejoramiento de Maiz y Trigo (CIMMYT), *1992/1993 World Wheat Facts and Trends* (Singapore: CIMMYT, 1993), pp. 43, 49. For Chile's 1982 and 1983 wheat imports, see *SALA* 29:841. The import total for 1981 is from United Nations, *International Trade Statistics Yearbook 1983*, vol. 1: Trade by Country (New York: United Nations, 1985), p. 171. For the 1981–83 cost of Chile's wheat imports, see United Nations, *International Trade Statistics Yearbook 1985*, vol. 2: Trade by Commodity (New York: United Nations, 1987), p. 19. For Chile's 1964–68 production and yield data, see FAO, *Production Yearbook 1969* (Rome: FAO, 1970), p. 38; for 1979, see FAO, *Production Yearbook 1980* (Rome: FAO, 1981), p. 96; for 1980–82, see FAO, *Production Yearbook 1982* (Rome: FAO, 1983), p. 108; and for 1983, see FAO, *Production Yearbook 1985* (Rome: FAO, 1986), p. 110.

11. FAO, *Production Yearbook 1994*, p. 68.

12. What follows is based on an interview with José Roberto Alvarado, INIA, 15 March 1990, pp. 1–6. Also see Roberto Alvarado, ed., *Manual de Producción de Arroz* (Talca: INIA, 1989).

13. FAO, *Production Yearbook 1987*, p. 118; and FAO, *Production Yearbook 1994*, p. 70.

14. What follows is based on notes from the Parral field trip, INIA, 16 March 1990, pp. 17–19.

15. What follows is based on an interview with Carlos Altmann Morán, INIA, 21 March 1990, pp. 30–33. Also see Carlos Altmann Morán, *G.T.T.: Un modelo Chileno para la Transferencia de Technología Agropecuaria* (Santiago: INIA, 1988) and Carlos Altmann Morán, *G.T.T.: La importancia de trabajar en grupo* (Santiago: INIA, 1989).

16. IRRI, *IRRI Rice Almanac 1993–1995* (Manila: IRRI, 1993), p. 129.

CHAPTER FOUR

1. For the 1994 population figures, see U.S. Bureau of the Census, *World Population Profile: 1994* (Washington, D.C.: Government Printing Office, 1994), p. A-6. On Colombia's rice production and per capita consumption, see International Rice Research Institute (IRRI), *IRRI Rice Almanac 1993–1995* (Manila: IRRI, 1993), pp. 51, 129, 130.

2. In 1970, almost 30 percent of Colombia's rice acreage was in IR8; see Grant M. Scobie and Rafael Posada T., *The Impact of High-Yielding Rice Varieties in Latin America with Special Emphasis on Colombia* (Cali: CIAT, 1977), pp. 15–16.

3. The acronym ICA stands for Instituto Colombiano Agropecuario.

4. See Robert F. Chandler Jr., *Rice in the Tropics: A Guide to the Development of National Programs* (Boulder, Colo.: Westview Press, 1979), pp. 129–31, and Scobie and Posada T., *Impact of High-Yielding Rice Varieties*, pp. 15–16.

5. The acronym FEDEARROZ stands for Federacíon de Arroz.

6. Inter-American Development Bank (IDB), *Economic and Social Progress in Latin America, 1986 Report* (Washington, D.C.: IDB, 1986), pp. 114–15; see also Grant M. Scobie and Rafael Posada T., "The Impact of Technological Change on Income Distribution: The Case of Rice in Colombia," *American Journal of Agricultural Economics* 60 (1978): 86–88.

7. Scobie and Posada, "Impact of Technological Change," p. 115. On these trends, also see Chandler, *Rice in the Tropics*, pp. 133–36, and Scobie and Posada T., *Impact of High-Yielding Rice Varieties*, p. 20.

8. What follows is based on an interview at La Libertad with Dario Leal, CIAT, 11 May 1989, pp. 14–17.

9. Comparable data can be found in IRRI, *Rice Almanac*, pp. 49, 51.

10. See my interview with Carlos Franco, CIAT, 10 May 1989, pp. 20–21.

11. For an analysis of where the germ plasm in released varieties originated, see Elicio P. Guimarães, Federico Cuevas-Pérez, and César P. Martínez, "Status of Rice Improvement in Latin America and the Caribbean," in *Rice in Latin America: Improvement, Management, and Marketing*, ed. Federico Cuevas-Pérez (Cali: CIAT, 1992), pp. 13–40.

12. On INGER Latin America, see the bulletin of CIAT's rice program, *Arroz en las Américas* 14 (December 1993): 10–11, and W. Ronnie Coffman et al., "A New Approach to Rice Improvement in Latin America," in Cuevas-Pérez, *Rice in Latin America*, pp. 34–35.

13. For the 1969–71 production data, see IRRI, *World Rice Statistics 1990* (Manila: IRRI, 1991), pp. 4–5; for 1989–91, see Food and Agricultural Organization of the United Nations (FAO), *Production Yearbook 1991* (Rome: FAO, 1992), p. 72; and for 1992–94, FAO, *Production Yearbook 1994* (Rome: FAO, 1995), p. 70; the averages subtract out U.S. production but include the Caribbean. For rice prices and the 1990 irrigated rice production figures, see CIAT, *CIAT On-Line* (August 1993): 1.

14. The list is from Guimarães, "Rice Improvement," p. 15.

15. IRRI, *World Rice Statistics 1990*, pp. 10–11 and 17; see also Iván D. Salas, "Arroz en Venezuela," *Arroz en las Américas* 12 (July 1991): 2–4 and Francisco Andrade and Carlos Monteverde, "Arroz en Ecuador," *Arroz en las Américas* 12 (July 1991): 5–6. For the 1992–94 figures, see FAO, *Production Yearbook 1994*, p. 70.

16. What follows is based on an interview with Luis Roberto Sanint, CIAT, 9 May 1989, pp. 3–5.

17. What follows is based on an interview with Robert Zeigler, CIAT, 9 May 1989, pp. 1–2. Zeigler left CIAT for IRRI in 1992.

18. Centro Internacional de Agricultura Tropical (CIAT), "Does International Agricultural Research Pay Its Dues?" *CIAT Press Release* (14 June 1992): 3.

19. IDB, *Economic and Social Progress*, p. 115.

20. What follows is based on field notes compiled at Santa Rosa; see CIAT, 11–12 May 1989, pp. 6–13.

21. See T. W. Mew, "Rice Diseases," in *Rice*, ed. Bor S. Luh, 2 vols. (New York: Van Nostrand Reinhold, 1991), 1:200–204.

22. Thomas R. Hargrove, "Know Your Enemy: Biotech Advance Holds Key to Defeating World's Top Rice Killer," *CIAT Press Release* (19 January 1993): 2–3. Also see R. S. Zeigler, S. A. Leong, and P. S. Teng, eds., *Rice Blast Disease* (Tucson: University of Arizona Press, 1994).

23. What follows is based on an interview with Ernesto Andrade at Semillano; see CIAT, 12 May 1989, pp. 37–41.

24. What follows is based on fieldnotes compiled at Puerto Lopez and interviews with Surapong Sarkarung, Bob Zeigler, and José Ignacio Sanz; see CIAT, 12–13 May 1989, pp. 28–35. Sarkarung left CIAT for IRRI in 1991.

25. CIAT, "El germoplasma africano trae diversidad genetica al arroz," *Arroz en las Américas* 13 (April 1992): 7. On the problem of genetic uniformity in centralized breeding, see Coffman et al., "New Approaches to Rice Improvement," pp. 31–37.

26. CIAT, "Savannas: Where the Grass Is Greener than Ever," *CIAT Press Release* (September 1992): 1–4.

27. On the importance of tropical forests and biodiversity, see the articles in E. O. Wilson, ed., *Biodiversity* (Washington, D.C.: National Academy Press, 1988), and Nigel J. H. Smith, J. T. Williams, Donald L. Plucknett, and Jennifer P. Talbot, *Tropical Forests and Their Crops* (Ithaca, N.Y.: Cornell University Press, 1992).

28. For a discussion of Brazil's policy in the Amazon, see Susanna Hecht and Alexander Cockburn, *Fate of the Forest* (New York: Harper, 1990), pp. 104–41. See also Kenneth Maxwell, "The Tragedy of the Amazon," *New York Review* (7 March 1991): 24–29, and Kenneth Maxwell, "The Mystery of Chico Mendes, *New York Review* (28 March 1991): 39–48.

29. What follows is based on the author's interview with José Toledo, CIAT, 10 March 1988, pp. 7–9.

30. What follows is based on the author's interview with Myles Fisher, CIAT, 21 June 1988, pp. 54–56.

31. What follows is based on field notes compiled by the author at Quilichao and an interview with Patricia Avila, CIAT, 23 June 1988, pp. 62–65.

32. See José M. Toledo, ed., *Andropogon gayanus: A Grass for Tropical Acid Soils* (Cali: CIAT, 1990).

33. See R. Schultze-Kraft and R. J. Clements, eds., *Centrosema: Biology, Agronomy, and Utilization* (Cali: CIAT, 1990).

34. The following is based on field notes compiled by the author at Mondomo and an interview with Raul Botero, CIAT, 23 June 1988, pp. 65–67.

35. Based on the author's interview with Reiner Schultze-Kraft, CIAT, 21 June 1988, pp. 57–59.

36. On leucaena, consult National Research Council, *Leucaena: Promising Forage and Tree Crop for the Tropics* (Washington, D.C.: National Academy Press, 1984), and Smith et al., *Tropical Forests*, pp. 316–24. Also see National Research Council, *Calliandra: A Versatile Small Tree for the Humid Tropics* (Washington, D.C.: Academy Press, 1983).

37. Based on the author's interview with Raúl Vera, CIAT, 21 June 1988, pp. 50–33.

38. Based on the author's interview with Carlos Seré, CIAT, 21 June 1988, pp. 59–61.

39. What follows is based on field notes compiled by the author at Florencia; see CIAT, 28 June–1 July 1988, pp. 76–78.

40. On the importance, organization, and management of small-scale dairy farming in developing countries, see Richard W. Matthewman, *Dairying* (London: Macmillan, 1993).

41. See Florencia field notes, pp. 73–75.

42. What follows is based on interviews with Jiro Gomez and Carlos Seré, CIAT, 28 June–1 July 1988, pp. 86–88, 73–74.

43. What follows is based on Florencia field notes, CIAT, 28 June–1 July 1988, pp. 81–86.

44. Guia Rural Magazine, "O mal da braquiaria," *Guia Rural* (December 1989): 23–25.

45. See J. Fisher, I. M. Rao, M. A. Ayarza, C. E. Lascano, J. I. Sang, R. J. Thomas, and R. R. Vera, "Carbon Storage by Introduced Deep-Rooted Grasses in the South American Savannas," *Nature* (15 September 1994): 236–38.

46. See CIAT, "Taming the Wild Peanut," *CIAT Press Release* (June 1993): 1–3.

47. What follows is based on field notes compiled on farms; see CIAT, 28 June to 1 July 1988, pp. 89–96.

48. See CIAT, *Arroz en las Américas* 12 (July 1991): 14–15.

49. See CIAT, "Growing Rice in Acid Soils," *CIAT Press Release* (April 1993): 1–4, and CIAT, "A New Rice-Pasture Farming System," *CIAT On-Line* (April 1993): 1.

50. John Madeley, "Green Shoots Appear on Colombia's Arid Savannahs," *Financial Times of London*, 15 May 1993, p. 20.

51. CIAT, *Annual Report 1992–1993: CIAT at the Threshold of Sustainable Development* (Cali: CIAT, 1993), pp. 12–13.

CHAPTER FIVE

1. On the difficulty of transferring the Asian model to Africa, see Paul Richards, *Coping with Hunger: Hazard and Experiment in an African Rice-Farming System* (London: Allen and Unwin, 1986), pp. 1–3, 15–27. On IR8 and for statistics on West African rice systems, see Akinwumi Adesina, "Economics of Rice Production in West Africa,"

West African Rice Development Association (WARDA), *State of the Arts Paper*, 1991, pp. 2, 9, 11, 15.

2. Adesina, "Production in West Africa," pp. 2–3, 81–82.

3. Richards, *Coping with Hunger*, pp. 22–27.

4. WARDA, "Rice Trends in sub-Saharan Africa," WARDA pamphlet, 1993, pp. 3, 4, 6, 7, 5; figures on West Africa's contribution to total sub-Saharan production excludes Madagascar.

5. For WARDA's recent work on glaberrima rices, see WARDA, *Annual Report 1992* (Bauaké: WARDA, 1992), pp. 30–31. Also see the bulletin of CIAT's rice program, *Arroz en las Américas* 16 (May 1995), p. 10.

6. On WARDA's research strategy, see Richards, *Coping with Hunger*, p. 26, and David Seckler, ed., *Agricultural Transformation in Africa* (Arlington, Va.: Winrock International, 1993), pp. 44–46. On the Sierra Leone project, see WARDA, *A Decade of Mangrove Swamp Rice Research* (Bouaké: WARDA, 1990), pp. 7–11, 21–23, 40–41, and Adesina, "Production in West Africa," pp. 14–15.

7. WARDA, *Mangrove Swamp Research*, pp. 21–23, 40–41.

8. On diffused light storage, see Robert H. Booth and Roy L. Shaw, *Principles of Potato Storage* (Lima: Centro Internacional de la Papa (CIP), 1981).

9. See CIAT, *CIAT Annual Report, 1987* (Cali: CIAT, 1988), pp. 49–52. These observations are also based on extensive interviews at CIAT in 1988 and 1989.

10. For some uses of cassava, see Gregory Scott et al., *Product Development of Root and Tuber Crops* (Cali: CIAT, 1993).

11. For data on budgets, see Consultative Group on International Agricultural Research (CGIAR), *1984 Annual Report* (Washington, D.C.: CGIAR Secretariat, 1985), pp. 64–65, and CGIAR, *1991 Annual Report* (Washington, D.C.: CGIAR Secretariat, 1992), p. 34.

12. The World Bank has published more than twenty-five *CGIAR Study Papers*. Each paper covers the CGIAR's work in a specific country or issues relevant to international agricultural research.

13. See Douglas Pachico and Eric Borbon, "Technical Change in Traditional Farm Agriculture: The Case of Beans in Costa Rica," *Agriculture, Administration, and Extension* 26 (1987): 65–74.

14. CIAT, *CIAT International* 14 (April 1995): 4–7.

15. For example, see Lester Brown, "Who Will Feed China?" *Worldwatch Magazine* (September–October 1994): 10–19.

16. See Technical Advisory Committee (TAC) of the Consultative Group on International Agricultural Research (CGIAR), "Investment in Rice Research in the CGIAR: A Global Perspective," TAC Secretariat Report, March 1993, pp. xiii, 3, 7, 27–28, and IRRI, *Annual Report 1992–1993: Rice in Crucial Environments* (Los Baños: IRRI, 1993), p. 2.

17. The new ideotypes for different rice environments are described in IRRI, *IRRI Towards 2000 and Beyond* (Manila: IRRI, 1989), pp. 35–42. Also see IRRI, *Crucial Environments*, p. 2, and B. S. Vergara et al., "Rationale for a Low-Tillering Rice Plant Type with High-Density Grains," in *Direct Seeded Flooded Rice in the Tropics, ed.* IRRI (Manila: IRRI, 1991), pp. 39–50. Also see Peter T. White, "Rice the Essential Harvest," *National Geographic* (May 1994): 56.

18. TAC, "Rice Research in the CGIAR," pp. 29–33; also see pp. 12–18.

19. Hiroshi Yamagata and Kunisuke Tanaka, "The Site of Synthesis and Accumulation of Rice Storage Proteins," *Plant Cell Physiology* 27 (1986): 135–45, and N. Mitsukawa and K. Tanaka, "Genetic Manipulation of Storage Protein in Rice," in *Rice Genetics II*, by IRRI (Manila: IRRI, 1991), pp. 503–10.

20. See Patrick Smith, "Letter from Tokyo," *New Yorker* (14 October 1991): 105, 108–9, and White, "Essential Harvest," p. 69.

21. Smith, "Letter from Tokyo," pp. 108–9, 113–14, and James Fallows, *Looking at the Sun* (New York: Pantheon, 1994), pp. 210–11. Also see Karel van Wolferen, *The Enigma of Japanese Power* (New York: Vintage, 1989), pp. 60–65.

22. For South Korea, see "The Vice of Rice," *Economist* 11 (December 1993): 75–76.

23. For U.S. milled-rice exports, see Food and Agricultural Organization (FAO) of the United Nations, FAO *Trade Yearbook 1993* (Rome: FAO, 1994), p. 96. For U.S. rough-rice production, see IRRI, *World Rice Statistics 1990* (Manila: IRRI, 1991), p. 5, and FAO, *FAO Production Yearbook 1994* (Rome: FAO, 1995), p. 70. To convert milled to rough rice for the U.S. case, multiply by 1.3; see IRRI, *World Rice Statistics 1990*, p. 315.

24. Compiled from tables in U.S. Bureau of the Census, *Statistical Abstract of the United States: 1994* (Washington, D.C.: Government Printing Office, 1994), pp. 681–83. Also see IRRI, *IRRI Rice Almanac 1993–1995* (Manila: IRRI, 1993), pp. 103, 105.

25. FAO, *Production Yearbook 1994*, pp. 70–71.

26. FAO, *Production Yearbook 1994*, p. 70.

27. For the size of China's rural population, see FAO, *Production Yearbook 1994*, p. 28. On irrigation projects in China, see Frank Leeming, *The Changing Geography of China* (Oxford: Blackwell, 1993), p. 17, and He Kang, "Rice Production in China," in *Impact of Science on Rice*, by IRRI (Manila: IRRI, 1985), p. 71. The figures on irrigated rice acreage are from IRRI, *Rice Almanac*, p. 48. For China's research system, see Shenggen Fan and Philip G. Pardey, *Agricultural Research in China: Its Institutional Development and Impact* (The Hague: International Service for National Agricultural Research, 1992), pp. 8, 34. On China's development of semidwarf rices, see Justin Yifu Lin, "The Nature and Impact of Hybrid Rice in China," in *Modern Rice Technology and Income Distribution in Asia*, ed. Cristina C. David and Keijiro Otsuka (Boulder, Colo.: Lynne Rienner Publishers, 1994), p. 375.

28. Robert F. Chandler Jr., *An Adventure in Applied Science: A History of the International Rice Research Institute* (1982; reprint, Manila: IRRI, 1992), pp. 184–86.

29. Dana G. Dalrymple, *Development and Spread of High-Yielding Rice Varieties in Developing Countries* (Washington, D.C.: Agency for International Development, 1986), p. 44.

30. Chandler, *Adventure*, p. 186.

31. On agricultural policy in China since 1979, see Leeming, *Geography of China*, pp. 73–111. On rural credit, see He Kang, "Production in China," p. 71. For developments in Chinese agricultural during the 1980s, see the articles in Bernhard Glaeser, ed., *Learning From China? Development and Environment in Third World Countries* (London: Allen & Unwin, 1987).

32. See the production index in Fan and Pardey, *Agricultural Research in China*, p. 14.

33. IRRI, *Rice Almanac*, p. 46.

34. Leeming, *Geography of China*, p. 75.

35. IRRI, *World Rice Statistics 1990*, p. 2.

36. IRRI, *Rice Almanac*, p. 46.

37. Kang, "Production in China," pp. 70–73. On chemical fertilizer use, see Lin, "Hybrid Rice in China," p. 375.

38. See White, "Essential Harvest," p. 65.

39. Randolph Barker, Robert Herdt, with Beth Rose, *The Rice Economy of Asia* (Washington, D.C.: Resources for the Future, 1985) p. 61; for the 1981 hybrid acreage in China, see p. 264. The current hybrid acreage is from White, "Essential Harvest," p. 64. Also see Lin, "Hybrid Rice in China," pp. 375–93.

40. IRRI, *Rice Almanac*, p. 47.

41. Chandler, *Rice in the Tropics*, pp. 101–44.

42. See A. Affandi, "The Rice Revolution in Indonesia: The Indonesian Experience in Increasing Rice Production," in IRRI, *Impact of Science*, pp. 27–39.

43. IRRI, *Rice Almanac*, p. 60, and Paul Heytens, "Rice Production Systems," in *Rice Policy in Indonesia*, ed. Scott Pearson, Walter Falcon, Paul Heytens, Eric Monke, and Rosamund Naylor (Ithaca, N.Y.: Cornell University Press, 1991), p. 38.

44. IRRI, *World Rice Statistics 1990*, pp. 2, 14, 34.

45. Compiled from tables in Scott Pearson, Rosamond Naylor, and Walter Falcon, "Recent Influences on Rice Production," in Scott Pearson et al., *Rice Policy in Indonesia*, pp. 18–19.

46. Affandi, "Rice Revolution," pp. 30–33.

47. Paul Heytens, "Technical Change in Wetland Rice Agriculture," in Scott Pearson et al., *Rice Policy in Indonesia*, p. 103.

48. Affandi, "Rice Revolution," pp. 32, 34–35.

49. Chandler, *Adventure*, p. 54 and IRRI, *Rice Almanac*, p. 59.

50. FAO, *Production Yearbook 1994*, p. 70.

51. IRRI, *Rice Almanac*, pp. 55–57; for the area planted to modern varieties, see pp. 58 and 130.

52. IRRI, *Rice Statistics 1990*, p. 31.

53. E. A. Siddiq, "Eastern Indian Holds the Key," *Hindu Survey of Indian Agriculture 1990* (Madras: N. Ravi, 1990), p. 45.

54. The figures are from an interview with R. S. Rana, director, National Bureau of Plant Genetic Resources, Indian Council for Agricultural Research, New Delhi, 22 July 1994. Also see the description of India's research system in Ishwar Chandra Mahapatra, Dev Raj Bhumbla, and Shriniwas Dattatraya Bokil, *India and the International Crops Research Institute for the Semi-Arid Tropics* (Washington, D.C.: World Bank, 1986), pp. 1–22.

55. See R. S. Rana, "India Takes the Lead in Conservation of Genetic Resources," *Indian Farming* (October 1993): 39–40.

56. For the 1994 percentage of India's economically active population engaged in agriculture, see FAO, *Production Yearbook 1994*, p. 28. The remaining figures are for 1991; see IRRI, *Rice Almanac*, p. 55.

57. See Siddiq, "Eastern India," pp. 45–49; Eastern India includes the following states: Assam, Bihar, eastern Madhya Pradesh, Orissa, eastern Uttar Pradesh, and West Bengal.

58. Siddiq, *Eastern India*, p. 49.

59. Based on field notes compiled at ICRISAT, 14 July 1994.

60. For 1988 figures see FAO, *Production Yearbook 1990* (Rome: FAO, 1991), p. 72; for 1990, see FAO, *Production Yearbook 1991* (Rome: FAO, 1992), p. 72; and for 1992–94, see FAO, *Production Yearbook 1994*, p. 70. For the 1995 estimate of China's rice production, see "China's Grain Harvest Up," *Jakarta Post*, 24 October 1995, p. 9. There is growing concern about China's food security; see Lester R. Brown, *Who Will Feed China? Wake Up Call for a Small Planet* (New York: Norton, 1995).

61. Consultative Group on International Agricultural Research (CGIAR), "New 'Super Rice' Will Help Feed Half-Billion More People," *CGIAR Press Release* (23 October 1994): 1–5. Also see IRRI, "A New Rice Plant With Higher Yield Potential," unpublished mimeographed pamphlet, pp. 1–6.

62. Letter from Gurdev Khush to the author, 25 May 1995. Also see CGIAR, "Super Rice," p. 3 and IRRI, "New Rice Plant," p. 5. The information on Harahap's breeding project is based on an interview in Bogor, Indonesia, with IRRI liaison scientist Cezar Mamaril, 16 November 1995.

63. See IRRI, *Annual Report 1993–1994: Filling the World's Rice Bowl* (Manila: IRRI, 1994), pp. 8, 11–16. Projects also included perennial rice for upland areas and shoring up Asia's rice-wheat rotations. On nitrogen fixation in rice, see S. K. Dutta and Charles Sloger, eds., *Biological Nitrogen Fixation Associated with Rice Production* (Washington, D.C.: Howard University Press, 1991), pp. 221–75.

64. Paul H. Connett, "The Disposable Society," in *Ecology, Economics, Ethics: The Broken Circle*, ed. F. Herbert Bormann and Stephen R. Kellert (New Haven, Conn.: Yale University Press, 1991), p. 100.

65. Cynthia Pollock, *Mining Urban Wastes: The Potential for Recycling*, Worldwatch Paper 96 (Washington, D.C.: 1987), p. 10.

INDEX

Sentence *Combining:

*Improving Student Writing without Formal Grammar Instruction

*no. 15 in a series of research reports sponsored by the NCTE Committee on Research

Sentence *Combining:

*Improving Student Writing without Formal Grammar Instruction

By Frank O'Hare
Florida State University

National Council of Teachers of English
1111 Kenyon Road, Urbana, Illinois 61801

Library of Congress Catalog Card Number: 72-95432.
ISBN: 0-8141-4339-3.
NCTE Stock Number: 43393.

Fifth Printing, May 1976

National Council of Teachers of English

Research Report No. 15

Literacy, or the lack of literate skills, overshadows and outweighs every other problem and need sensed by educators and clients of the schools. The broad base of opinion to this effect is reflected in the highest priorities of current educational legislation and planning. For this reason, I was particularly interested to learn, prior to reading the manuscript for this book, that Frank O'Hare's work involved an instructional technique for use in teaching *writing* that was at least potentially capable of yielding results that would profoundly alter the current instructional practices of the writing curriculum. My reading of O'Hare's manuscript confirmed this description.

The instructional approach in the O'Hare study is called *sentence-combining*, a type of pedagogy involving extensive, sequenced practice of specially formulated print-based exercises through which a student is said to acquire dexterity in writing complex sentence structures. On its face, the sentence-combining technique has a solid foundation in research. The main ideas, though original in configuration, are supported by the work of several leading linguists and, indirectly, by the work of many behavioral scientists over a period of decades. One of the crucial *linguistic* notions here is that written English is a dialect distinct from spoken English, from which it would appear to follow that an effective pedagogy should be based upon language-learning techniques. Another notion is that the linguistic mechanisms of sentence generation are extremely dynamic, from which follows the possibility, indeed the actuality in the sentence-combining method, of devising learning activities in which the linguistic processes of sentence generation can be simulated by the student. The basic *psychological* ingredient has to do with an apparent fact about learning whereby complex skills are most readily learned when they are broken down into smaller component subskills, as when in the sentence-combining method a student matures in his linguistic ability in written English through a succession of quasi-generative learning experiences in sentence building. The methodology appears to have application as an instructional strategy at many levels of training in writing, from the elementary grades on.

O'Hare's contribution has been to identify the practice of sentence-combining as the probable cause of the positive effects that have been observed in a series of experiments in which sentence-combining activities were present, but not the exclusive elements of the treatment. For historical reasons, the sentence-combining technique arose within the context of a debate on the relevance of formal grammar instruction (in this instance, transformational grammar) to the acquisition of measurable writing skills. The force of O'Hare's work, which reports impressive positive effects for the exclusive use of sentence-combining, is to render the entire issue academic, at least with respect to the short-term goal of finding curricular and instructional solutions to the problem of illiteracy in writing.

One should bear in mind that O'Hare's experiment does not have laboratory characteristics, and although O'Hare himself is highly qualified by experience and training, the test data cannot be taken as conclusive proof, nor can they be fairly interpreted without reservations. Still, O'Hare has provided the first major test of sentence-combining methodology in a relatively pure form and, while important questions remain to be answered, I can think of no line of research in the area of writing that holds greater promise for effective curricular change than further exploration of sentence-combining as a pedagogy.

Peter S. Rosenbaum
For the Committee on Research

TABLE OF CONTENTS

LIST OF TABLES

ACKNOWLEDGMENTS

I wish to thank the members of my dissertation committee, John Simmons, chairman, Roy O'Donnell, Garrett Foster, and James Preu, for the helpful advice they gave me during my work on this study. My special thanks to Roy O'Donnell and Kellogg Hunt for their assistance with the problems that arose in connection with the grammatical analysis of the data; to Gerald vanBelle for his creative work with the statistical analysis of the data, especially the instruments he devised to assess overall writing quality; to James Barnes, who so ably taught two of the classes in the experiment; and to the eight experienced English teachers who judged the quality of the student writing.

John Mellon of Harvard University generously allowed me to use and adapt many of the sentence-combining exercises from his text *Our Sentences and Their Grammar*. For this I am truly grateful.

I also wish to thank Thomas Devine and Peter Rosenbaum of the NCTE Committee on Research and James McCrimmon of Florida State University, who helped me revise this manuscript.

F. O'H.

Tallahassee, Florida

INTRODUCTION

Despite Marshall McLuhan's timely warning to society in general and to educators in particular that we are at the end of the Gutenberg era, the age of writing and of printed materials, and that the electronic "nonwriting" age is upon us, educators remain convinced of the importance of writing as a humane, perhaps the most humane, skill developed by man. Written records have enabled man to pass down through the centuries his discoveries, his frustrations, and his aspirations. The eloquence of Cicero, the simplicity of the Sermon on the Mount, the wisdom of the Bhagavad Gita, all would be lost to us had we not devised the means to put them on paper.

The English-speaking community has given English teachers the responsibility of teaching people writing, the putting-words-down-on-paper skill. And English teachers who have been writing for a very long time have come up with a bewildering variety of "right" ways to teach writing. There are almost as many theories as there are theoreticians. Even more perplexing, although many of these theories make good sense, each in turn has been, if not refuted, at least called to question by contradictory evidence. After an exhaustive study of writing research Braddock (1963) concluded that

> Today's research in composition, taken as a whole, may be compared to chemical research as it emerged from the period of alchemy: some terms are being defined usefully, a number of procedures are being refined, but the field as a whole is laced with dreams, prejudices and makeshift operations. (p. 5)

At least since Aristotle the search has been on for an all-embracing theory of rhetoric or composition or plain writing. This metatheory would assign to their proper places and in their proper degrees such components of writing as ideas, organization, style, voice, tone, vocabulary; it would reconcile differences, confirm similarities, answer all our questions. Despite the fact that some impressive attempts have been made in recent times to formulate such a theory—for example, the works of I. A. Richards and Kenneth Burke—no completely satisfactory metatheory has appeared: opposites remain unreconciled, doubts unresolved.

Teachers of writing, however, are less interested in composition theory than in the practical implications of any given theory. Confronted daily with the task of improving student writing, these teachers cannot afford to wait until a satisfactory metatheory emerges from research. For them the crucial question is always a practical one: Will it make my students better writers?

The present study is not designed to test any proposed metatheory of composition. It has its eye, rather, on composition students in the English classroom, and, as a consequence, its aims are much more limited, much more specific. Interested in the possibilities of altering and improving students' writing behavior, this study seeks answers to the following questions:

1. Would seventh graders who practiced a new kind of sentence-combining exercise that was in no way dependent on their formal knowledge of a grammar write compositions that could be described as syntactically different from those written by students quite similar to them in ability who were not exposed to such sentence-combining practice?
2. If there were syntactic differences in their writing, could these differences be called differences in maturity?
3. Would the students who practiced the sentence combining write compositions that would be judged better in overall quality?
4. What would be the curricular implications of these findings?

Although this study could be described as being in the tradition of previous linguistic research on the relationship between grammar study and improvement in writing, it is not a grammar-based study. Indeed, it was an examination of recent linguistic studies of the relationship between grammar and writing that led this researcher to hypothesize that sentence-combining practice need in no way be dependent on formal knowledge of a grammar, traditional or transformational.

This study does, however, rely on transformational theory. The sentence-combining exercises written out by the students are entirely dependent on a theory of generative grammar. Equally important to this study were the recent transformationally oriented studies of Hunt and O'Donnell on the development of syntactic maturity. This researcher simply felt that, although a knowledge of transformational—or for that matter, traditional—grammar is an indispensable tool for the researcher and a potentially useful tool for the teacher of English, there was no justification for assuming that it would help students write better. However, the deliberate elimination of generative-transformational

grammar study from this experiment and the systematic exclusion of grammatical terminology from the entire experiment must in no way be construed as a rejection of grammar study *per se*. The large and very interesting question as to whether grammar should be studied in schools at all will not be dealt with in this study.

Recent studies dealing with the relationship between a certain kind of language study and writing are examined in Chapter One. The first part of Chapter Two demonstrates that normal growth in syntactic maturity can be measured in quantifiable terms. The second part of the chapter both describes and suggests a rationale for sentence-combining practice that is in no way dependent on students' formal knowledge of a grammar.

Chapter Three discusses the design and the procedures used in this investigation, including the hypotheses to be tested, the research design, the subjects, the independent, dependent, and extraneous variables, the experimental and control treatments, and the measurement and analytic procedures.

In Chapter Four the results of the analysis of the data are both presented and discussed.

The final chapter contains the conclusions of this study, the theoretical and practical implications of the conclusions for the teaching of writing, and, finally, some suggestions for further study.

CHAPTER 1

RECENT RESEARCH ON GRAMMAR STUDY AND WRITING

With the publication of *Syntactic Structures* in 1957, Noam Chomsky revolutionized grammatical theory. Subsequent refinement of his generative-transformational theory by Lees (1960, 1961), Chomsky himself (1965), and others has led to a general acceptance of transformational theory as an efficient method of formulating "the most economical and coherent system of explicit rules adequate to characterize all the grammatically well-formed sentences possible in a particular language" (O'Donnell, Griffin, and Norris, 1967, p. 15). Chomsky and the other transformationalists' demonstration of the superiority of certain aspects of generative over those of traditional grammar led Meckel (1963), in a survey of the effects of the teaching of grammar on writing, to observe that "much of the earlier research on teaching grammar must be regarded as no longer of great significance outside the period in educational history which it represents" (p. 982).

While acknowledging generally the truth in what Meckel has said, it is nevertheless interesting to examine some of the studies completed before Chomsky which concerned themselves with the relationship between formal grammar study and writing because that history is a curious one indeed. Study after study tested the hypothesis that there was a positive relationship between the study of grammar and some aspect or other of composition. Result after result denied this hypothesis. Many of the findings either clearly indicated, or at least strongly suggested, that the study of grammar not only did not have the desired result, but that there also resulted some undesirable side effects. Braddock (1963), in a review of formal grammar and its effect on writing, declared that

> In view of the widespread agreement of research studies based upon many types of students and teachers, the conclusion can be stated in strong and unqualified terms: the teaching of formal grammar has a negligible or, because it usually displaces some instruction and practice in actual composition, even a harmful effect on the improvement of writing. (pp. 37–38)

Subsequent reviews of research by Bateman and Zidonis (1964) and

Mellon (1965*a*, 1967) corroborated Braddock's statement. The reader might ask, "Why 'curious' then?" The history of this extensive research is a curiosity simply because *so much* research has been conducted on this question. Why were English researchers so persistent? Why didn't they recognize that there was indeed no relationship between formal grammar study and writing?

The answer is, perhaps, a surprisingly obvious one. English teachers must have *instinctively* felt that somehow, somewhere, someone would find the connection that they "knew" was there. For over a century teachers had been teaching grammar and expecting, indeed assuming, that it would help their students write better.

Further proof of this almost mystical faith in the efficacy of grammar study can be found in the nature of the studies produced by linguistic researchers after the publication of Chomsky's *Syntactic Structures*. Instead of abandoning this line of investigation altogether, these researchers immediately set out to examine the claim that exposure to generative-transformational grammar would improve students' writing. And it wasn't long before their optimism paid off. Between 1964 and 1968 there appeared several studies whose results indicated that the transformational grammar approach did have an effect on student writing.

Bateman and Zidonis published a study which claimed that a knowledge of generative grammar enabled students to increase significantly the proportion of well-formed sentences they wrote and to increase the complexity without sacrificing the grammaticality of their sentences. In 1968 this researcher was a doctoral student at the Florida State University in the Experienced Teacher Fellowship Program and participated in some lively discussions of the Bateman and Zidonis study and of a similar kind of study by John Mellon. These discussions led the present researcher to the idea that perhaps the Bateman and Zidonis study was successful only because of the sentence manipulation their students had performed, and to wonder whether Mellon's grammar study had hindered his students in any way.

Although Mellon was at pains to differentiate his study from that of Bateman and Zidonis, the two studies proved to be remarkably similar. They were, as Mellon (1969) claimed of the Bateman and Zidonis study, "the [first experiments] in the entire canon of grammar and writing research that explicitly [advanced] a sentence-structure hypothesis" (p. 10). Both exposed their students to the study of a generative-transformational grammar. Both were interested in the possibility that error reduction would result from their experimental treatments. Bateman and Zidonis concluded that a knowledge of gen-

erative grammar could enable students to reduce the occurrences of errors in their writing. Mellon originally planned an analysis of error incidence but abandoned the project as too time consuming and expensive. In their findings both studies claimed that their students wrote sentences that were syntactically more complex or mature. And they both relied heavily on generative grammar for their various analyses.

Of course, there are obvious differences between these two studies, which a brief appraisal of Bateman and Zidonis and a more detailed look at Mellon should make clear.

The Bateman and Zidonis Study

The study conducted by Bateman and Zidonis (1964) was a landmark in the history of research investigating the effect of grammar on writing because it hypothesized that the study of a transformational grammar would affect the structure of the sentences students wrote. As with any pioneer study, it should be looked upon as a product of its time and a reflection of the state of knowledge of that period. In 1962–63, when this study was conceived and planned, Chomsky's generative-transformational grammar was relatively new. Extravagant claims were being made for its "generative" capabilities by overenthusiastic supporters; equally strenuous denunciations flowed from traditional grammarians who perhaps felt threatened. A further difficulty was the complex prose style of Noam Chomsky, who used a sort of "linguistic shorthand" frequently couched in the language of mathematics to "clarify" his ideas. Not exactly ideal fare for the average English teacher, who typically abandons the study of mathematics at an early age. Feelings were high, misconceptions rife, acrimony bitter.

It is not surprising, then, that when Bateman and Zidonis discovered that their experimental group had reduced the incidence of errors and at the same time had employed more mature sentence structures, they concluded that it was a result of their students' knowledge of transformational grammar. Generative grammarians stood vindicated. Or so it appeared.

No useful purpose would be served by examining the Bateman-Zidonis study in detail, but since several criticisms leveled at it by Mellon deserve some attention, a brief description of some aspects of the study might prove useful.

The investigators selected the ninth grade of the University School of Ohio State University and randomly assigned the students to two classes. Over a two-year period the experimental class studied the

"regular curriculum" and specially prepared materials from the area of generative grammar. Six pre-test and post-test compositions were collected from both sections during the first three months of the first year and the last three months of the second year. The investigators reported the sentences according to whether they contained errors or not. They followed this by calculating the mean "structural complexity scores" for each of the two sentence types. The structural complexity of a sentence was derived by adding one to the number of transformations each sentence contained. Forty-six transformational rules were listed by the investigators and used to identify the transformational history of each sentence.

Bateman and Zidonis reported that their experimental students' study of transformational grammar enabled them to increase significantly the proportion of well-formed sentences they wrote and to reduce the occurrence of errors in their writing. The increase in average structural complexity scores for well-formed sentences was 3.79 for the control class and 9.32 for the experimental—which, of course, represented an increase of over five transformations per sentence. Interesting results despite the fact that the greatest changes in the experimental group were made by only four students. Mellon rightly questioned whether analysis of variance was the appropriate statistic here. Nevertheless, he seemed to ignore an indisputable fact: the experimental students did write sentences of greater complexity. Four students comprised approximately one-fifth of the experimental population. Although some of Mellon's criticisms are well founded, others are, perhaps, a little too severe. Mellon took the investigators to task for not utilizing the findings of Kellogg Hunt, ignoring the fact that the Bateman-Zidonis study, a two-year enterprise, was completed and published in the same year as Hunt's study (1964). Also questionable is the severity of Mellon's reaction to the investigators' description of what the control class studied. Mellon was correct, of course, when he suggested that more information should have been given than the following:

> Each class studied what would be considered the regular curriculum at the school with this exception: the experimental class studied materials specially adapted by the investigators from the area of generative grammar. (1964, p. 10)

> In each class, improvement of pupil writing was one of the major objectives. The classes differed only in content: no formal grammar was studied in the control class; the grammatical content described in Chapters 2 and 3 was studied by the pupils in the experimental class. (1964, p. 117)

Some sort of outline of the English courses that the experimental and control students were exposed to would no doubt have been useful. However, "regular curriculum" does give the reader some idea of what went on. Surely any unusual subject matter or technique employed would either have been avoided or reported in some detail. It does seem that Mellon reacted too strongly when he declared, "Surely this is a major oversight in such a study" (1969, p. 13).

Mellon's final criticism of the Bateman-Zidonis study is sound. He claimed that the hypothesis of the entire experiment was based on a line of argument which was difficult to accept rationally, and rightly took Bateman and Zidonis' study to task for claiming that "pupils must be taught a system that accounts for well-formed sentences before they can be expected to produce more of such sentences themselves" (1964, p. 3). Abundant research has demonstrated that young children have already mastered a very large proportion of the structures of English before they get to school and quickly learn to handle the remainder in elementary school. However, this whole question of language development will be discussed in detail later.

Although the hypothesis of the Bateman and Zidonis study was based on a questionable assumption and had certain methodological problems, it is nevertheless a significant study. Being wise after the event is a favorite practice of researchers, and "if only I had . . ." a common cry. This is within the nature of man. Their study was a pioneering one. That others would follow was inevitable. That they would profit from mistakes and oversights is within the nature of the discipline. The significance of this study lies in the discovery that students who study transformational grammar end up writing sentences that have fewer errors and are more complex syntactically than students who do not. That is significant indeed.

The Mellon Study

Mellon's purpose was to find out whether students who were exposed to what he called "transformational sentence-combining practice" would significantly increase their normal rate of growth in syntactic ability. The results indicated that the students who were exposed to the treatment showed statistically significant increases in what Mellon called "syntactic fluency."

Mellon's study was a reaction to the study of Bateman and Zidonis and showed signs of having profited from Bateman and Zidonis' experiences. Mellon was not as interested in the possible corrective function of his sentence-combining practice, the error-reduction effect

of so many previous studies, but concentrated on the other important aspect of the Bateman-Zidonis study, the increase they found in syntactic complexity. He rejected the Bateman-Zidonis claim that the learning of grammatical rules *per se* could lead to improvement in student writing or that these rules could be applied in any *conscious* manner by the writer. Mellon suggested that it was the sentence-combining practice and not the study of the grammar that had an effect on the students' writing behavior, a point which will be examined later.

Mellon's experimental population consisted of 247 white native-American middle-class seventh grade students in twelve classes in four schools in the Boston area. The schools were chosen to represent urban, suburban, and private education.

There were three separate treatments. Five experimental classes studied a year-long course in transformational grammar that included a large amount of sentence-combining practice. Five control classes studied a course in traditional grammar. Two placebo classes studied no grammar at all but had extra lessons in literature and composition, but no additional writing assignments. All twelve classes studied the regular English program for their particular schools.

The writing sample at each test time consisted of nine compositions, each written in one class period during the first four and last four weeks of school. Mellon selected, for each student before and after the treatment, the first ten T-units from each composition that the student wrote, ninety T-units in all at each test time.* Mellon adapted the T-unit which Hunt (1965) described in his study *Grammatical Structures Written at Three Grade Levels.* (Hunt's and Mellon's T-units will be discussed later.) The main dependent variables in the study were twelve factors of syntactic fluency, including T-unit length, subordination-coordination ratio, the number of nominal and "relative" clauses and phrases (which included adverbial clauses of time, place, and manner), clustered modification, and depth of embedding.

Comparison of pre- and post-test results indicated that the experimental group showed increases in all twelve factors and that the gains were significant at or beyond the .01 level of confidence. Mellon also compared the increases achieved by his group with normative data from the Hunt study. Hunt had established normal per year growth for nominal clauses and phrases and relative clauses, phrases, and words. Mellon found that his experimental group showed from 2.1 to

*A T-unit consists of a principal clause and any subordinate clause or nonclausal structure attached to or embedded in it.

3.5 years of growth on these same factors, while his control group failed to show even one year's growth. Mellon's hypothesis that the writing of the experimental group would show a significant increase in syntactic fluency was substantiated. The English teaching world was justifiably impressed by these findings.

Although the Mellon study is a substantive piece of innovative research, it does pose a number of rather interesting problems. The first problem is an unusual one. Which of the Mellon reports is being discussed? There are two reports, one published in 1967 by Harvard University and the other in 1969 by NCTE. In the present discussion reference will be made to the NCTE report. However, statements from the Epilogue of the 1969 report will be considered later because, in this researcher's opinion, the Epilogue constitutes something akin to a change of mind on Mellon's part and, indeed, lends some measure of support to the hypothesis of the present study, which was outlined in the spring of 1969.

A critical problem facing anyone examining the Mellon study is the question as to what exactly constitutes transformational sentence-combining practice. Does "transformational" mean simply that the students' practice was based on the *researcher's* knowledge of deep and surface structure which led him to construct the combining problems in kernel form? Or does "transformational" imply the *student's* knowledge of generative grammar? Examination of *Our Sentences and Their Grammar* (1965*b*), Mellon's 162-page experimental text, reveals that he taught generative grammatical concepts all the way through. For example, on pages 137–138 the concept of pre-noun modifier is presented and the students are encouraged to be able to identify adjective phrases, participles, passive participles, and participial compounds. Although the student is "taught" these concepts in one and a half pages, he never uses them and indeed never encounters them again until they appear as part of a rather formidable list—for seventh graders, at any rate—in the last lesson (p. 157).

Mellon's experimental treatment demanded three things of his students: (1) that they learn transformational rules like T:rel, T:gerund, T:der-NP, T:infin, which they had to apply in the combining practice; (2) that they learn concepts like passive infinitive phrase, appositive noun phrase, participial compound, etc., which they were never asked to apply consciously in the combining practice; and, most important, (3) that they learn a quite difficult set of grammatical rules (how well we are never told). Mellon described his grammar as elementary. The present researcher finds it difficult to believe that seventh graders

would find it easy to memorize and/or conceptualize such theoretical constructs as:

T:gerund = NP + AUX + VERB + NP → NP + S + VERB + ING + OF + NP.

T:infin = NP + AUX + VERB + remainder → FOR + NP + TO + VERB + remainder.

T:der-NP = NP + AUX + VERB + NP → NP + VERB +
URE + OF + NP.
ANCE
MENT
TION
AL

The last example is a particularly intriguing one because a student who assiduously followed the rule for T:der-NP, which is summarized on page 106 of the experimental text, would have found, when writing out problem 3 on page 107, that the rule did not work. "Simmons published the experiment" cannot be changed by literal adherence to Mellon's rule to "Simmons' publication of the experiment. . . ." Indeed, on the same page (107) there are five more examples, *none* of which will give the student the proper answer if he follows Mellon's T:der-NP rule.

Analysis of many of these rules forced this investigator, who had taught English to seventh graders for about ten years, to conclude that many were too difficult and that Mellon's average and below-average students were perhaps using the examples and largely ignoring the theoretical apparatus when they wrote out their sentences.

One is left, then, with an insoluble problem. Was it the study of this particular transformational grammar that led to the syntactic gains made by Mellon's experimental group? Or was it the combining practice only that led to these increases? Or was it the interaction of the grammar study and the combining practice? The design of Mellon's study does not permit this question to be answered because, as previously mentioned, he taught a grammar that was only partially utilitarian and exposed the students to combining practice too.

There is evidence that this grammar was quite difficult. Perhaps Mellon's experimental group would have shown greater increases in syntactic fluency if the grammar studied had been easier. The grammar studied may have inhibited some students and in some way counteracted possible gains. Again, the study's design excludes the answer to this question. Mellon could perhaps have had a fourth group study the grammar alone and write out a limited number of illustrative sentences to clarify the particular concept being studied.

Mellon did not mention whether the experimental students were

tested as to their relative knowledge of the grammatical concepts studied. Was there a relationship between any of the twelve factors of syntactic fluency and the students' relative knowledge of the grammar studied? Mellon declared that

> The chief purpose of this course was neither to rectify the student's language behavior nor to facilitate the sentence-combining practice. Rather, it was to present to junior high school students, in an obviously introductory manner, an elementary transformational grammar describing the language competence they and all other speakers already possess. As with contemporary studies in other curriculums, the main justification for this course was given in terms of the experiences and learnings generated by the inquiry it occasioned. (1969, p. 27)

If the chief purpose was to teach an elementary grammar, it would seem desirable to have tested the students' knowledge of that grammar. The general impression one gets from an examination of the experimental text is that grammatical concepts are being *mentioned* but not thoroughly taught. For example, highly complex branching diagrams appear on page 116, and yet the student is never asked to construct one. On pages 127 and 128 six very difficult grammatical terms are introduced, illustrated but not defined, and then summarized—all on scarcely more than a page. They never appear again until the overall review, where they are simply listed. It would be a rare seventh grade student indeed who could learn the terms participial phrase, passive participial phrase, infinitive phrase, passive infinitive phrase, prepositional phrase, and appositive noun phrase by being shown only one example of each.

Mellon also asserted that

> . . . as an activity designed to reinforce and further illustrate transformations earlier learned by the student, the problem-solving practice was considered an integral part of the grammar course and may be viewed in this light quite without regard for its possible effects upon syntactic fluency. Its role was very much like that of the straightforward exercises in formula application which are employed, for example, in modern school algebra. (1969, p. 27)

An examination of *Our Sentences and Their Grammar,* especially of the second half, would suggest that this is not entirely the case. Indeed, Mellon added thirty daily five-minute problems which were not included in the text and which are further proof that the combining practice was not simply used for illustrative purposes. In fairness to Mellon it should be pointed out that in the Epilogue he wrote in 1969, he freely admitted to this charge, agreeing that there would be no need for so many examples, especially any involving multiple embedding.

While agreeing with Mellon that his study should have been presented in what he described as an "a-rhetorical" setting, the present researcher is not at all convinced that sentence combining should *remain* "a-rhetorical." Mellon continually insisted on the "a-rhetorical" nature of his study, when there is limited but indisputable evidence that his treatment was not consistently rhetoric-free. For example, in the Preview Lesson of the treatment, Mellon says to his students,

> Now that you are beginning junior high school, you will be devoting a great deal of time to developing your writing skills. Will the study of grammar help you to write better? No one really knows the answer to this question. But there are several reasons for thinking that it may. . . . By the end of the school year, you should be writing sentences more skillfully than you do now. (1965*b*, p. 2)

This is a clear assurance that grammar study should help them to write better. Later, while commenting on an exercise where the student has to embed about eight statements or kernels into a given sentence, "Yesterday we read over those manuscripts," Mellon says,

> Even though we have not used all the modifiers possible, you may feel that we have used too many. That is, we may have chosen too many additional things to say about the "manuscripts" in our main-clause statement. . . . thus, you should choose these details carefully and try to build effective noun modification. The use of *appropriate* added detail often is the difference between interesting writing and ordinary lackluster work. For example, here is a "story" that consists of three sentences whose nouns are unmodified:
>
> > A girl set out for a picnic into the woods.
> > There she met a wolf.
> > The wolf joined her for a lunch.
>
> Now we shall add several insert sentences to each of the main-clause sentences given above. Notice how these inserts provide descriptive detail and give our "story" a recognizable tone. (1965*b*, pp. 145–146)

On page 141 he says, "You will find that the repeated modification of a single noun sounds quite natural."

It is obvious that there is, perhaps inevitably, a rhetorical tone to these statements, an implied or explicit exhortation to the student to use these devices in his own writing. There is a concern with how "it" sounds. Doesn't this imply an audience critically reading or listening? The question of what Mellon means by "a-rhetorical" and its implications for the present study will be taken up again in the last chapter.

Finally, there is the rather interesting question as to when an enhanced syntactic maturity would become discernible to the general reader. Bateman and Zidonis ignored this issue, and Mellon was satisfied that no harmful side effects appeared. If sentence-combining

practice is designed to make students write sentences that are more mature syntactically, it seems reasonable to assume that at some point this syntactic difference would show qualitatively. The results of Mellon's quality evaluation were disappointing. The *control* group was judged on post-tests to have written compositions that were significantly better than those written by the experimental group. Mellon attributed these results to the small sample size and/or the effect of one especially talented control teacher.

Would a larger, more reliable sample have favored the experimental group? Mellon, whose experimental group increased their T-unit length by approximately 1.2 words per T-unit, suggested that this increase is not sufficient to become noticeable even to an experienced grader. Two rather interesting questions arise quite naturally from this observation: (1) At approximately what point would an experienced grader recognize that there were syntactic differences in the students' writing? and (2) Would these syntactic differences influence the grader's evaluation of the students' writing?

The Bateman-Zidonis and Mellon experiments, then, exposed students to the study of transformational grammar, and both studies showed that their experimental groups wrote sentences which were syntactically more mature. Bateman and Zidonis assumed that it was a knowledge of generative grammar and its application that enabled their experimental subjects to write differently. Mellon called to question such an assumption, claiming that it was the combining practice, not the grammar, that enabled his students to write differently. But the design of the Mellon experiment makes it impossible to ascertain whether the study of transformational grammar had a positive or negative or no effect on the students' syntactic development.

Therefore, although there is at least some doubt as to what exactly caused their students to write differently, there can be no doubt that both in the Bateman and Zidonis study and in the Mellon study the experimental treatments significantly altered the writing behavior of students exposed to them.

Audio-Lingual Studies

Another group of linguistic researchers who have altered the syntactic writing behavior of their students are the advocates of the audio-lingual or oral-drill technique, which has, of course, been used in the teaching of foreign languages for a number of years. Three rather interesting audio-lingual experiments by Ney (1966), Raub (1966), and Miller and Ney (1968) have been undertaken with gen-

erally significant results. Because the methodology in all of these experiments was, according to Ney, "basically very similar since they followed the model of the pilot project with seventh graders which was reported in the *English Journal* by Ney" (pp. 2–3), no useful purpose would be served in analyzing each study in detail.

In all three studies the experimental treatment was designed "to condition the students to use sentences of predetermined syntactic types through verbal manipulation of representative sentences from oral cues" (Ney, 1968, p. 2). The three studies also included written exercises which were related to the oral exercises in order to effect transfer of training from speech to writing. In all but one of the experiments the progress of the students was measured by having the students write for an unspecified time about a film which they had just been shown. The sentences in these compositions were classified by type and counted according to techniques devised by Hunt (1965) and O'Donnell, et al. (1967). The experimenters were interested primarily in finding answers to two questions: (1) Did the experimental classes write more of the structures that they had been conditioned to use than their respective control classes? and (2) Were the experimental classes' sentences syntactically more mature?

In these studies the pre- and post-test film was the same, to control for the possible influence of subject matter on the syntactic structures the students might use. Since students readily use the syntactic patterns they hear in a film, they were shown, with one exception, films without narrative or dialogue. Ney summed up the test results by declaring that,

> In the three experiments in which pretests and posttests were given, improvement in the form of a greater frequency of occurrence on the posttests of the structures practiced was always measureable although it did not always reach a level at which the experiment was statistically significant. (1968, p. 4)

It is obvious, then, that audio-lingual techniques cause some change in student writing behavior.

The Miller and Ney Study

The last of the three experiments briefly described here, the Miller and Ney study (1968), was the most interesting for the purposes of the present study. The Miller and Ney study compared the performance of a fourth grade experimental class which was exposed for one year, September to June, to regular oral practice in manipulating syntactic structures with a fourth grade control class that had regular lessons in reading and composition.

The experimental class was exposed to the treatment four days a week during thirty-seven periods of from thirty to forty minutes and in the second half of the experiment, two days per week during thirty periods averaging forty to fifty minutes. Typically, students were asked to repeat two cue sentences which were written on the blackboard, for example,

> The boy put the old man down.
> The boy was very tired.

The teacher would then read the sentences in a combined form, e.g., "The boy, who was very tired, put the old man down." Then the students would perform choral reading of this sentence. These were reinforced by ten similarly structured combined sentences which were practiced orally by the class. Review exercises were also constructed which contrasted the differing sentences studied. The students were also given practice in writing out correct sentences when the teacher read sets of cue sentences.

Generally speaking, the treatment was designed to produce three types of sentences:

1. sentences with *who* and *which* adjective clauses:
   A. He looked at the boy. The boy came out of the river.
   B. He looked at the boy who came out of the river.
2. sentences with initial and final adverb clauses:
   A. The princess couldn't be married. She was too proud.
   B. The princess couldn't be married because she was too proud.
   B. Because she was too proud, the princess couldn't be married.
3. sentences with subject and predicate nominals derived from the deep structure:
   A. Something disturbed the king. The princess talked.
   B. The talking of the princess disturbed the king.

After the oral practice the students participated in choral reading from various textbooks, from rewrites of Mark Twain's work, and from folk tales written for foreign students. This kept the lessons interesting, gave the students additional practice, and provided a linguistic context for the language exercises.

Both the experimental and control groups showed an increase from the first pre-test in the structures which were taught, but only the gain shown by the experimental group was statistically significant. The results from the second post-test indicated that the experimental group was using the structures practiced far more frequently than the control group and that the experimental gain was significant at the

.001 level of confidence. It was felt that the gain achieved by the control group was directly attributable to the effect of the narration on the first film. The experimental group also wrote a greater number of words than the control group, and the increase in experimental output was statistically significant. Analysis of variance indicated that on both the incidence of structures practiced and number of words produced at the second post-test the difference in performance was in favor of the experimental group and significant at the .001 level of confidence.

The Miller and Ney study also measured the length of multi-clause and single-clause T-units and compared the scores of the experimental and control groups by analysis of variance. On these measures the experimental group showed a generally greater improvement than the control group, the most impressive gain being made in the number of words in multi-clause T-units from the second pre-test to the second post-test. The experimental group wrote just over twice the number of words in multi-clause T-units on the second post-test as they had done on the second pre-test. In comparison with the control group the experimental group wrote fewer simple sentences and proportionately more complex sentences.

Summary

It is clear from the evidence of the studies surveyed in this chapter that there is, contrary to the findings of the many traditional-grammar writing studies, a very real connection between a certain kind of language study and writing. Recent experiments have effected change in student writing behavior. The experimental groups wrote sentences that were syntactically more mature. Oral and written drills undoubtedly made a difference. Although it is at least questionable whether it was a knowledge of generative grammar that led Bateman and Zidonis' students to write more mature sentences, it is not unreasonable to assume that something in their experimental treatment must have caused those students to write more maturely. And lastly, although the design of Mellon's study does not provide an obvious basis for his conclusion that it was the sentence-combining practice that made the difference, common sense tells us that Mellon was probably right. Indeed, Mellon's multiple embedding of kernel sentences is the most promising technique yet developed for utilizing transformational theory. However, the unanswered question remains: What effect did the grammar study have on Mellon's experimental group?

CHAPTER 2

SYNTACTIC MATURITY AND SENTENCE COMBINING

English teachers have always been aware that on the average younger children do not write as well as do older children, that high school students write better than elementary students, and, of course, that educated adults write better than high school students. Included in this judgment were decisions not only about the ideas, organization, vocabulary, and spelling used by these groups, but also about the style. It was obvious to these teachers that the more mature writer somehow put his ideas down differently. His sentences were generally longer. He put more into his sentences by lengthening his independent clauses and by using more subordination. These sentences were usually more complex, fancier, harder to read. English teachers called the older student's style more mature. And if his style failed to please them, it might be called immature or choppy. Comments like "Your sentences lack flow" or advice like "Try to write more naturally" are commonplace on students' papers. "Your sentence structure lacks maturity" is perhaps an accurate but not very helpful piece of advice to give to a student. He might even ask his teacher what was meant. And what to do about it.

In the present study maturity of sentence structure will be defined in a statistical sense as the range of the sentence types found in samples of the students' writing, and it will usually be referred to as "syntactic maturity." Generally speaking, English teachers have been able not only to distinguish between elementary and high school student writing, but also to identify normal stages of development in student writing as typical of a particular grade range. Confronted by a composition written by a fifth grader, an experienced teacher could describe it as syntactically mature or immature or normal. The trouble was and is that teachers often disagree. Differences of opinion about something as vague as style would be inevitable. What was needed was some kind of objective measure that would confirm the intuitions teachers feel about maturity of sentence structure and describe the features that constitute syntactic maturity in quantifiable terms. If quantification could be satisfactorily accomplished, then

normal stages of syntactic development could readily be identified and objectively verified.

The advantages of such measures for the English teacher are obvious. Students who are not displaying "normal" growth could be quickly identified and given remedial instruction. But what about the student who is developing "normally"? Is this the best that he can do? Students who come from a home environment that has rich cultural and linguistic experiences would predictably be above average in syntactic maturity. Since variations in syntactic maturity are indisputable and since normal growth is really only a way of saying average growth, it seems reasonable to assume that this normal rate of growth could be accelerated or retarded under certain treatment conditions. This study thus directs itself to the question of whether sentence-combining practice will enhance the normal growth of syntactic maturity.

Language Development Studies

The many studies of language development that have been published have been critically reviewed by Heider and Heider (1940), McCarthy (1954), and more recently by Carroll (1960), Erwin and Miller (1963), Mellon (1965*a*), and O'Donnell, et al. (1967). There is therefore no advantage to be gained by reviewing the literature again.

Traditionally, observations on language development or syntactic maturity have identified the lengthening of sentences and increased use of subordinate clauses as indicators of progress toward a mature style. More recently, several normative studies have further specified the syntactic characteristics that distinguish the writing of older from that of younger writers. Two of the more important recent studies on language development are those done by Hunt (1964, 1965) and O'Donnell, et al. (1967). Because of limited time and resources which necessitated his performing all the syntactic segmenting and counting himself, the present researcher was anxious to find an economical, efficient, and reliable measure or measures of syntactic maturity. Although Mellon had investigated twenty variables and had found all but two of them to be significant at the .05 and more often beyond the .01 level of confidence, many of these measures seemed to be highly redundant. The findings of the Hunt and O'Donnell studies suggested a reasonable compromise.

Hunt (1965) investigated 1000-word samples of the free writing of school children in grades 4, 8, and 12 and the writing of skilled adults who published in *The Atlantic* and *Harper's*. Hunt introduced a new

measure called a "minimal terminable" unit or "T-unit," which was a refinement of Loban's (1961, 1963) "communication unit." The T-unit is one main clause plus any subordinate clause or nonclausal structure that is attached to or embedded in it. The experimental population in the present study consisted of seventh graders, and anyone who has taught seventh graders knows how notoriously forgetful they can be when it comes to remembering to put down something as remote from their daily concerns as a period at the end of a sentence. Could groups of words be called a sentence when they displayed all the characteristics of a sentence, including being followed by a capital, if they did not in fact terminate with a period? Also, some students put periods where we would put commas. What to do about that? The solution was simple: ignore the sentence and concentrate on a more reliable and more objective measure, the T-unit. We're interested, then, among other things, in the incidence of clauses and T-units.

Hunt discovered that as students get older they tend to write longer clauses and that skilled adults carry that tendency further. Maturing children write more clauses per T-unit, but skilled adults do not carry that tendency much further than do twelfth graders. As they get older, these children write longer T-units, and skilled adults carry that tendency even further because they tend to write lengthier clauses.

Hunt also discovered that for grades 4, 8, and 12 the best of these indexes of syntactic maturity is T-unit length. Second best is clause length; third best is clauses per T-unit. When the writing of skilled adults is included in the sample, there is only one difference. Words per clause becomes as significant an index of syntactic maturity as words per T-unit.

In the O'Donnell study (1967) the investigators sampled the speech of thirty children in kindergarten and thirty in grades 1, 2, 3, 5, and 7. They also took writing samples from the students in grades 3, 5, and 7. After viewing two eight-minute films with the sound turned off so that the narrator's language would not influence their language production, the children were asked to tell the story of the film privately to an interviewer and to answer certain questions related to the narrative.

O'Donnell, using Hunt's T-unit to segment the student's output, found that at every grade the average length of the T-unit increased. The number of clauses per T-unit also increased with the child's age. Although O'Donnell did not report on the number of words per clause, this figure can be calculated from his data on T-unit length and clauses per T-unit. The clause length figures calculated from O'Donnell's data are similar to those of Hunt, showing an increase at each grade level (Hunt, 1970, p. 9).

Even a casual examination of Table 1 reveals that for words per T-unit, clauses per T-unit, and words per clause there is generally a steady increase moving up the grades for the combined figures of O'Donnell and Hunt through grade 7. Hunt's figures for grades 8, 12, and beyond indicate a continuation of this steady increment.

Table 1

Words per T-Unit, Clauses per T-Unit, Words per Clause

Grade Level	3	4	5	7	8	12	Superior Adults
Words/T-Unit							
O'Donnell	7.67		9.34	9.99			
Hunt		8.51			11.34	14.4	20.3
Clauses/T-Unit							
O'Donnell	1.18		1.27	1.30			
Hunt		1.29			1.42	1.68	1.74
Words/Clause							
O'Donnell	6.5		7.4	7.7			
Hunt		6.6			8.1	8.6	11.5

Source—Adapted from Hunt (1970). Based on data reported by Hunt (1965) and by O'Donnell, et al. (1967).

Both Hunt and O'Donnell also investigated the number of sentence-combining transformations used by their experimental subjects and discovered that this number increased as the subjects got older. Hunt found that older writers, especially skilled adults, use a much larger number of transformations per T-unit and per clause and concluded that this explained the fact that clauses, especially those of skilled adults, increased in length with maturity.

In examining subordinate clauses, Hunt reported that the most important developmental trend was an increase in adjective clauses, which more than doubled in frequency, the percentage increase being slightly greater during the second half of the time span. The number of adjective clauses per T-unit for grades 4, 8, and 12 was .045, .090, and .16, respectively. This represented an almost fourfold increase. Hunt concluded by suggesting that the increase in number of adjective clauses was most important as an index of maturity. His superior adults used .25 per main clause, which was more than the number used by his twelfth graders. Hunt then declared that, for adjective clauses,

> the rate of increase from one of the four groups to the next is remarkably steady, and also rather dramatically large. Over the three grades

> (4, 8, 12) the increase is nearly fourfold, but if we include the fourth group (superior adults) the increase is more than fivefold. The likelihood that a fourth grader will embed an adjective clause somewhere in a T-unit is only 1 in 20. The likelihood that a superior adult will do so is 1 in 4. (1965, p. 90)

Hunt reported also that noun clauses significantly increased, although their overall percentage increase was about half that shown by adjective clauses.

Hunt's skilled adults wrote T-units 40 percent longer than did his twelfth graders. The older writers not only wrote more subordinate clauses per main clause, especially adjective clauses, they also wrote longer clauses, which, of course, combined with the greater number of clauses, accounted for their writing longer T-units. The O'Donnell study stressed the importance of T-unit length as the most effective single measure of syntactic maturity:

> This investigation supports the findings by Hunt (1964, 1965) that when fairly extensive samples of children's language are obtained, the mean length of T-units has special claim to consideration as a simple, objective, valid indicator of development in syntactic control. (1967, pp. 98–99)

O'Donnell also suggested that the enormously time-consuming process of counting the kind and depth of every sentence-combining transformation might perhaps be regarded as redundant when he stated that

> The readily performed calculation of mean lengths of T-units, however, appears to give a close approximation to results of the more complicated accounting of sentence-combining transformations. (1967, p. 98)

While agreeing that older writers employ different sentence structures than do younger writers, the English teacher might have observed that older writers simply deal with different subject matter. Perhaps it was the constraints of this subject matter that accounted for the more mature syntax used by older writers. The Hunt and O'Donnell studies that have already been examined dealt with two different kinds of free writing, and free writing rendered this question unanswerable.

An experiment conducted by Hunt (1970) was designed to find out whether students, differing in age and maturity level, and adults would display different levels of syntactic maturity when confronted with the *same* subject matter. These students would say the same thing, because each was given a set of extremely simple sentences to combine and instructed to utilize all of the information they contained. When a writer added any idea that was not contained in the original thirty-two sentences, the whole sentence was deleted.

The instrument used in the study was developed by Roy O'Donnell of the Florida State University. It is a passage containing thirty-two simple sentences, which the students were instructed to "write in a better way." The instrument was administered to over a thousand students, almost exclusively white, in grades 4, 6, 8, 10, and 12 in Tallahassee's public schools. From each grade level fifty students were chosen to represent "something close to a normal distribution of academic ability" and from the scores for each student's writing means were computed for each grade and for the high-, middle-, and low-ability groups in each grade. Although Hunt used a number of new measures, and got especially interesting results with what he called structures less than a predicate and less than a clause, his findings relating to the number of embedding transformations and to clause and T-unit length were of particular interest to the present study.

Although all the writers were required to say the same thing in Hunt's experiment, the older writers displayed superior syntactic manipulative ability. Their sentences were affected by their syntactic maturity. Older writers tended to use a wider variety of transformations when reducing inputs to less than a predicate. They wrote longer clauses and longer T-units. Interestingly, the trends indicated in Hunt's 1970 study are the same as those shown in his and O'Donnell's studies of free writing.

Of particular concern to the present study were Hunt's findings that the number of embeddings of kernels correlated highly with clause length and that syntactic maturity consisted chiefly in the ability to make many embedments per clause. Hunt demonstrated that syntactic maturity involves a manipulative skill that is, in some sense, independent of subject matter. Even when the older writer added no more information, he still wrote more words per T-unit and more words per clause. He displayed more syntactic maturity.

The research reviewed in the present study has shown that as the child matures, he tends to embed more sentences, which results in an increase in clause and T-unit length in his writing. Perhaps these increases can be attributed to his cognitive development. Or perhaps they are the result of his imitating the more mature styles that he encounters in his reading and in conversation at school. Whatever the reason, there is clearly a developmental trend. Therefore, since they tend to increase with age and are indicative of a developing linguistic maturity, the syntactic characteristics outlined here would appear to be efficient criteria for describing syntactic maturity.

Pattern Practice and Modeled Writing

In his study Mellon considered several "grammar-related activities" that might lead to "syntactic fluency" and dismissed them in turn. These activities were modeled writing, pattern practice, applied transformational rules, and traditional sentence parsing. Since the present researcher agrees with Mellon's able rejection of applied transformation rules and traditional sentence parsing, nothing more need be said here about them.

Pattern practice requires students to write sentences in accordance with a series of grammatical commands. Mellon dismissed this activity because he claimed that it forces the student to search for "pointless content" and thus distracts him "from the very thing to which he is supposed to be attending, namely, the given pattern" (1969, p. 21). While agreeing that there is a possibility of students being distracted when engaged in this activity, the present researcher cannot agree that these are grounds for the abandonment of such an activity. An imaginative teacher could so structure such an assignment as to make the students' search for meaningful content interesting *per se*. For example, one could imagine a series of exercises in which the student is given one, two, then three blanks to fill in in a partially written sentence whose deep structure might contain six or seven kernels. Mellon seemed not to take into account modeled writing that would involve sentence cues, thus ignoring a long series of studies which utilize such cues in their audio-lingual approach to learning a foreign language and learning English as a second language. These studies, which will be discussed subsequently, surely involve some kind of modeled-writing pattern practice. And they have proven quite successful.

These are, perhaps, peripheral issues, but in developing a rationale for transformational sentence-combining practice, Mellon defended two assumptions which may be called to question and which are of prime concern to the present study. He repeatedly claimed that sentence-combining practice must be "a-rhetorical" in nature. He also asserted that

> the chief purpose of [the grammar] course was neither to rectify the student's language behavior nor facilitate the sentence-combining practice. . . . the problem-solving practice was considered an integral part of the grammar course and may be viewed in this light quite without regard for its possible effects upon syntactic fluency. (1969, p. 27)

The present study will take issue with both of these assertions in due course. However, it does agree with Mellon's other conditions for

efficient sentence-combining practice, namely, that the final behavior elicited from the student should be the production of a "fully formed statement whose structure is predetermined and characteristic of mature expression" and that the content must be given in a format that will facilitate the production of the desired sentence.

Sentence-Combining Practice Not Dependent on Formal Knowledge of a Grammar

However one may disagree with certain aspects of Mellon's rationale for transformational sentence-combining practice, one is faced with a hard fact. It worked. Mellon claimed that the combining practice was an integral part of the grammar study. The present researcher questions this claim. Mellon also declared that the combining practice must be a-rhetorical. Although this assertion is at least a matter for conjecture, and indeed it will be called to question later in this study, there can be no disputing the fact that Mellon's transformational sentence-combining practice was conducted in a largely a-rhetorical setting.

The rationale for the present study grew out of experiences the present researcher had in his senior secondary classes in his native Scotland, where he lived until he had completed an M.A. from Glasgow University. A favorite activity in Scottish English classes was an exercise which supplied the student with perhaps six or seven kernel-like statements and directed him to "write all of these as a compound-complex sentence with two adjective clauses, an adverb phrase, and two adverbial clauses, one of concession and one of place." Suppose, as often happened in this researcher's case, the wretched student did not know either what a compound-complex sentence was or what an adverbial clause of concession was. In a classroom environment where physical punishment for unsatisfactory work was an everyday occurrence and its avoidance an attractive alternative, the student would simply work with what he did know and use his intuition for what remained. And he was quite often successful in coming up with the correct answers. (Perhaps the "paddle" hasn't been sufficiently investigated as a sentence-combining stimulus!)

Years later this researcher chanced on the Mellon treatment and immediately set about solving the sentence-combining problems without benefit of Mellon's grammatical signals. It was a fairly difficult enterprise at first. But it wasn't long before "(T:rel)" became "who" or "which" or "that" and, more difficult, "(T:fact)" and "(T:exp)" combined to become "it . . . that." After that the hunt was on. Could all

of these grammar labels be eliminated and a series of practical "little helps" substituted? The advantages of this approach were legion. The students wouldn't have to study a grammar that, for seventh graders at least, was really quite difficult. This would eliminate any possible adverse effects of the grammar study on students who simply could not understand the *theoretical* constructs which, as has been previously shown, were not thoroughly taught in Mellon's treatment, but rather mentioned *en passant*. The simplicity of the instructions would allow the student to deal, unhampered, with concepts which language-development research has demonstrated he already has mastered. The complexity of the problems would not be reduced but could be faced square on. Success might breed success. Students don't like traditional grammar study, not only because it is boring, but also because many of them simply cannot do it. With grammar gone, the full potential of Mellon's kernel-embedding system might be realized.

It does not seem unreasonable to assume that, when they were writing out the sentence-combining problems, at least some of Mellon's students may have gone through an experience similar to this researcher's teenage experiences in Scotland. Although they might have had only a sketchy knowledge of the grammatical concepts, they probably had recourse to their own practical linguistic experience or, more likely, they flicked back in the text and found a combining problem, already solved, exemplifying a similar problem.

The sentence-combining practice in this study, while freeing the student from the distraction of seeking meaningful content himself, would give him systematic and controlled experience in the production of sentences which were more mature than those he would ordinarily write. He could give his undivided attention to the actual process of transforming by addition and deletion without worrying about grammatical theory.

John Mellon very graciously gave the present researcher permission to use and change his sentence-combining problems as he thought fit. The present study retained at least 95 percent of Mellon's sentences so that comparisons of the results could be made. Occasionally Boston area street names, sports arenas, etc., were changed to their equivalents in Tallahassee. Very little else was changed. Since some of the thirty extra five-minute daily exercises that Mellon's students worked on were not available, this researcher substituted an equivalent number of similar kinds of combining problems.

However, the present study incorporated very important changes in Mellon's format, and these changes were so important as to alter the very nature of the activities. This study is a replication of Mellon's

only in the sense that the students' final product was a series of similar sentences.

Mellon's students were exposed to the study of a transformational grammar throughout the year. They learned a series of transformational concepts which apparently facilitated the solution of the sentence-combining problems. To illustrate the form of his transformational sentence-combining problems Mellon used the following example that would appear in about the seventh month of the grammar course:

> *Problem:*
> The children clearly must have wondered SOMETHING.
> The bombings had orphaned the children.
> SOMETHING was humanly possible somehow. (T:wh)
> Their conquerors pretended SOMETHING. (T:infin–T:exp)
> Chewing gum and smiles might compensate for the losses. (T:fact)
> The losses were heartbreaking.
> They had so recently sustained the losses.
>
> *Write-out:*
> (Here the student writes the fully formed sentence.) (1969, p. 22)

Mellon explained the process like this:

> Briefly, the right hand indentations show how the embedding is to proceed. The first sentence is always the main clause. The sentence or sentences immediately beneath it and spaced one place to the right are to be embedded therein, and so on down the list of successively right-spaced sentences. The capitalized word "SOMETHING" indicates an open nominal position, repeated nouns signal relativization, and parenthetical items are abbreviated transformational directions where necessary. (1969, p. 23)

The present study abandoned entirely the formal study of grammar because grammar study was not needed. What was needed was a series of simple, consistent, practical, and efficient signals designed for the sole purpose of facilitating the sentence-combining operations. They had to be easy to understand and easy to use.

Right-hand indentations were abandoned because they also were not needed. Students could perform the combining operations without them. The capitalized word *SOMETHING*, which indicated an open nominal position, was retained because students found it easy to understand and very helpful. Students had trouble with the concept of a repeated noun signaling relativization. Instead, this researcher simply underlined the relative words that would be retained in the write-out. Students found this particularly helpful because all they had to do was to get rid of anything that was not underlined in that line.

A similar example should help the reader to compare and contrast

the two methods of facilitating the sentence-combining practice.

> The children clearly must have wondered SOMETHING.
> The bombings had orphaned the children. (WHOM)
> SOMETHING was humanly possible somehow. (WHY)
> Their conquerors pretended SOMETHING. (IT–FOR–TO)
> Chewing gum and smiles might compensate for the losses. (THAT)
> The losses were *heartbreaking.**
> They had so recently sustained the losses. (WHICH)

The student, in both Mellon's and the present study, was instructed to move down the sentences, combining them as he went, into one sentence. If successful, he wrote it out as follows:

> The children whom the bombing had orphaned clearly must have wondered how it was humanly possible for their conquerors to pretend that chewing gum and smiles might compensate for the heartbreaking losses which they had so recently sustained.

Although additional illustrations are provided in Appendix A, there is really no substitute for a thorough examination of the gradual building up process that was an integral part of the present study.

The system employed in this study has some advantages over Mellon's system. A student doesn't have as many abbreviated grammatical instructions to keep in mind. For example, in Mellon's system, "(T:wh)" could mean "who," "what," "when," "where," or "why." The new system, as the reader can see from the example above, specifically tells the student to use "why," or "whom," or "which."

As every English teacher knows, students very often don't know when to use "who" or "whom." The repeated nominal doesn't help, but the new system virtually guarantees the correct form by telling the student which one to use. "(T:infin)" and "(T:exp)" are instructions for Mellon's students to use both the infinitive transformation and the expletive transformation. The new system's "(IT–FOR–TO)" takes the guesswork out and allows the student to confront the real issue—the embedding problem itself.

The new system is demonstrably easier because it focuses on the needs of the student. If the purpose of a sentence-combining instruction system is to facilitate the sentence-combining operation, then that is precisely what it ought to do. It should anticipate where the *student* will be likely to encounter a problem. And then it should help the student solve that problem right when he needs the help.

In preexperiment trials the students who worked on these problems found them quite interesting for several reasons. The most important

---

*The italicized word was underlined in the experimental text.

reason was that they were easy to do. They gave students confidence with sentence manipulation. A student had to test his answers against his own sense of grammaticality. Seeing in his mind's eye sentences "click together," as one student put it, as he moved down the kernels was a positive reinforcement of the sentence-combining process. Students were also impressed with the maturity of the write-out sentences and often claimed credit for them, referring to them as "my sentences." There was every reason to believe, then, that the students in the experiment would find these problems challenging and interesting.

The greatest attraction for both teacher and pupils of the system of sentence-combining practice described here is, of course, that it does not necessitate the study of a grammar, traditional or transformational. The English teacher who simply "doesn't like grammar" can use this system. Also, many English teachers, although attracted to transformational grammar in theory, are repelled by some of its very complicated rules, especially its tree diagrams. Others maintain that, because generative grammar is in its infancy and could quickly become obsolete, learning its many complicated rules could be a waste of their time. And, of course, many English teachers, troubled by grammar's demonstrable lack of utilitarian value, nevertheless feel that grammar study is an important part of human knowledge. All of these teachers can use the present system because it avoids the negative aspects of grammar study altogether.

Grammar study is in disrepute at the present time largely because it has failed to help students write any better. It has occupied the center of language study in the classroom, and many people, including some grammarians, think that this is regrettable. In a lively article called "Linguistics and the Pursuit of Relevance" Neil Postman suggested that the grammarian and his works should be placed "at the distant periphery of language study, not at its center" and that "the primary goal in language teaching is to help students to increase their competence to use and understand language" (1967, p. 1162). He denounced the idea that language should be studied "for its own sake," and then asserted that a very important goal in the teaching of English is "helping students to manage their lives more effectively by increasing their control over language" (p. 1162). Postman quoted I. A. Richards, "America's greatest living linguist," in an article in *The New York Review of Books*. Writing of the general failure of teachers to make the study of language relevant and useful, Richards said:

> It was *not* the badness of the grammar descriptions which caused the failure but a simpler and deeper mistake: learning how to describe a

> language is not at all the same as learning how to *use* it with power and discernment. In point of fact, current efforts by English teachers to use transformational grammar far too often result in glib manipulation of nomenclature—just as of old—and play with "tree diagrams" without bringing any improved understanding of what sentences do or how they do it. (Postman, 1967, p. 1162)

Richards and Postman clearly identified the problem. English teachers have been too concerned with how language works and not sufficiently concerned with developing ways to help students use their language. Obsessed with theory, they have ignored practice. Michael Scriven, in a speech delivered to the College of Education faculty at the Florida State University, called this tendency the "academic fallacy" and gave as an absurd example the view that "one can't swim without having a satisfactory theory of hydrostatics, hydrodynamics and the physiology of immersed activity." He suggested that the serious examples of the "academic fallacy" are built into our curriculum; "at the college level, for example, the laughable idea that symbolic logic is a significant aid to reasoning skill in any substantial field, that French grammar has something important to contribute to French-using skills. . . ." Aren't English teachers also guilty of the academic fallacy when they stress *why* to the detriment of *how*?

Richards, Scriven, and Postman all stressed the importance of the *use* one makes of a skill. And that is precisely what sentence-combining practice is designed to do—to make students better able to handle English sentences. Of course, there was no suggestion here that the students would write in their free writing sentences as long as those they practiced. What was postulated in this study was that there would be a sort of "rub-off" effect from sentence-combining practice with multiple embeddings which would lead to greater syntactic maturity in free writing. Football players practice hundreds of plays many times so that at the right time, in the right situation, a dozen or so of these moves will have become both appropriate and habitual. So also with sentence combining. Only some of the operations should become habitual.

A further attraction of sentence-combining practice is that it forces the student, as he embeds the given kernels into the main statement, to keep longer and longer discourse in his head. Practice at memorizing and reproducing these longer sentences may help him develop a skill which two researchers at least have claimed is characteristic of increasing cognitive maturity. Harrell (1957) discovered that younger children write shorter sentences than they speak, and his evidence suggests that older children are better than younger children at learning

to keep sentences of increasing length in their heads while writing them out on paper. Support for Harrell's contention can be found in the theory of "chunking" which Miller (1956) developed. Miller suggested that as the mind matures it develops a more sophisticated ability to organize complex information. According to Hunt, this developing ability would explain why children, as they mature, produce and receive more complex sentences (1970, p. 58). Miller's explanation of how the memory span, which is a fixed number of chunks, can handle additional information by building larger chunks containing more information than before is an attractive one and would appear to support the kind of sentence-combining manipulations being advanced in the present study. Miller declared that

> In the jargon of communication theory, this process would be called recoding. The input is given in a code that contains many chunks with few bits per chunk. The operator recodes the input into another code that contains fewer chunks with more bits per chunk. There are many ways to do this recoding but probably the simplest is to group the input events, apply a new name to the group, and then remember the new name rather than the original input events. (1956, p. 93)

Obviously Miller's description of the recoding process is very similar to the series of operations demanded by the sentence-combining practice in the present study. "Bits of information" are very like kernels which have to be embedded, and "chunks" are similar to relative and nominal kernel embeddings. The long T-unit is, therefore, the result of the reduction and embedment of bits of information into chunks which naturally become larger.

The case for this study's sentence-combining practice is a strong one both from a practical and a theoretical standpoint. It should facilitate syntactic skills already possessed by "training" the memory and increasing the cognitive "chunking" ability of the students. The system is simple and can be learned by the average English teacher in several inservice sessions. Because it demands an acceptant, non–error-oriented environment that accentuates the positive, students should find it easy to do and relatively interesting. Few students should make many mistakes.

Curricular Assumptions

The present experiment presented an interesting curricular dilemma. Although the sentence-combining practice was presented in an a-rhetorical setting because of design requirements, this experimenter is not at all convinced that that is the only or even the best method

of presenting sentence-combining practice. In order to isolate the effect of sentence-combining practice, the exercises were given to the students in a carefully structured and almost entirely a-rhetorical setting. Great pains were taken to avoid conditioning students to favor complex syntactic expression in their actual composition classes.

At least one-half of the regular composition course for all students consisted of the students writing a journal each week, with the stress in this part of the course on encouraging students to develop a sense of personal worth. The uniqueness of the individual and the importance of the day-to-day happenings in the life of that individual were of paramount importance. Sentence structure, punctuation, spelling, etc. were largely ignored. The major responsibility of the teacher was to react as another human being, as sympathetically as possible, to the searching, the joys and disappointments, the uncertainties, the probing of the individual student. Content was all important. The major objective was to get the students to increase the flow of their writing and in so doing to improve their self-concepts.

The sentence-combining problems were never referred to in the composition class. In fact, they were systematically avoided. Had the sentence-combining practice been presented in concert with or as an integral part of the composition instruction, major problems of interpretation of the results would doubtless have arisen. In such a situation, if the students' writing behavior had changed significantly, perhaps the change could be attributed to the unique effect of sentence-combining practice *and* composition instruction.

This is not to say that sentence-combining practice *ought* to take place in an a-rhetorical setting. Indeed, the present researcher strongly believes that sentence-combining practice has very real attractiveness when considered as an integral part of composition instruction because (1) it has such a direct bearing on the generally neglected question of style and (2) it has potential usefulness for the student who is revising a paper which has been condemned for an immature or choppy style, quite without regard to its effectiveness in the present experiment.

Mellon (1969) believed that secondary school English should consist of "three autonomous component subjects—literature, composition, and linguistics" and that "linguistics and composition are separate subjects in pursuit of separate goals." This position is at best questionable. Since the Anglo-American Seminar on the Teaching of English held at Dartmouth College in 1966 there has been a discernible movement among English educators away from the tripartite division of the English curriculum, and the present researcher is certainly in favor of

such a trend. However, no useful purpose would be served by embarking on a lengthy discussion of the relative merits of these two seemingly opposing views because the design of the present experiment necessitated the isolation of sentence-combining practice from the rest of the curriculum so that its effect on student writing could be directly measured. The whole question as to whether sentence combining must be, as Mellon claims, "a-rhetorical" (p. 20), whether it has "nothing whatsoever to do with . . . the teaching of writing" (p. 81), and whether it is "not a program of composition or rhetoric" (p. 74), will be returned to in the last chapter of this study, where the general curricular implications for the sentence-combining practice described here will be discussed at length.

Summary

The first part of this chapter demonstrated that growth in syntactic maturity can be measured in quantifiable terms and that the six factors utilized in the present study constitute a reasonable measure of syntactic maturity.

The second part of the chapter both described and developed a rationale for the present study's system of sentence-combining practice. Sentence-combining practice that was in no way dependent on the students' formal knowledge of transformational grammar should increase the normal rate of growth of syntactic maturity in the students' free writing. Practice with intensive sentence manipulation that involved multiple embedding of kernels supplied in advance and the final development and production of sentences considerably more mature than normally written and spoken by such students should result in an enhanced cognitive ability to produce sentences that are syntactically more mature.

Non–error-oriented, grammar-study-free, and wholly dependent on each individual student's inherent sense of grammaticality, the sentence-combining practice virtually guaranteed student success, and success should produce a positive, acceptant classroom atmosphere that, in stressing the spirit of inquiry, would encourage syntactic experimentation and build confidence. The dais might disappear; the student as syntactic authority take over. At least a part of linguistics could more nearly become "student-centered"—certainly a desirable curricular development.

CHAPTER 3

DESIGN AND PROCEDURES

The overall plan of this study was to test whether sentence-combining practice that was in no way dependent on the students' formal knowledge of transformational grammar would increase the normal rate of growth of syntactic maturity in the students' free writing in an experiment at the seventh grade level over a period of eight months. In this experiment subjects were randomly assigned to the experimental and control groups. Samples taken from pre- and post-treatment compositions were used as a basis for determining syntactic maturity. The amount of growth experienced by an experimental group was compared with that of an equivalent control group and where possible, with the normative data reported by Hunt (1965), O'Donnell, et al. (1967), and Mellon (1969). With the obvious exception of the sentence-combining practice, the experimental group was exposed to the same kinds of units as the control group. The experimental units were simply shorter. Both the experimental and control group wrote the same number of compositions, plays, speeches, etc.

An evaluation of the overall writing quality of a subsample of the total sample's writing output was also undertaken to determine whether the actual growth in syntactic maturity experienced by the experimental group would influence the judgments of a group of eight experienced English teachers called upon to compare the overall quality of matched pairs of a sample of the experimental and control compositions.

Hypotheses

The study was designed to test the following two major hypotheses for significance at the .05 level:

1. The experimental group, which was exposed to the sentence-combining practice, will score significantly higher on the six factors of syntactic maturity than the control group, which was not exposed to the sentence-combining practice.
2. The experimental group will write compositions that will be

judged by eight experienced English teachers as significantly superior in overall quality to the compositions written by the control group.

This study also tested for possible interaction effects of teacher, sex, and ability as measured by IQ and pre-test scores on words per T-unit on the syntactic maturity of the students' writing.

Since the subsample of the compositions written consisted of fifteen pairs of narrative and fifteen pairs of descriptive writing, this study tested whether the eight teachers judged the narrative and descriptive compositions differently. Also of interest was whether the teachers as a group agreed in their evaluations of these compositions.

Research Design

Design of the Study. The experiment was designed to include two experimental and two control classes to which students were randomly assigned. The pre-test–post-test control group design described by Campbell and Stanley (1968, p. 13) was utilized. The design took thc following form:

R O X O
R O   O

Subjects. All of the eighty-three seventh grade students at the Florida State University High School were included in the study. These students were within the normal seventh grade range of 12 to 13 years and had IQ scores ranging from 76 to 143 with an average of 111.6. Thirteen percent of the students were black. There were forty-three boys and forty girls in this predominantly middle class population. There were forty-one students in the experimental group and forty-two in the control group.

Variables. The following outline summarizes the independent, dependent, and extraneous variables of this experiment.

I. Independent.
   Methods and materials: teaching a regular curriculum in English versus sentence-combining practice and a shortened version of this regular curriculum.

II. Dependent.
   A. Six factors of syntactic maturity.
      1. Words per T-unit.
      2. Clauses per T-unit.
      3. Words per clause.
      4. Noun clauses per 100 T-units.

5. Adverb clauses per 100 T-units.
6. Adjective clauses per 100 T-units.

B. A single qualitative judgment, based on the factors of ideas, organization, style, vocabulary, and sentence structure, made concerning which of two compositions, one experimental and one control, was superior. (The superior composition was assigned a score of *one* and the other composition was assigned a score of *zero*. The compositions had been matched according to the subjects' sex and IQ.)

III. Extraneous.

A. Language experiences of the subjects outside their English classes.

B. The two teachers who each taught an experimental and control class.

Procedures

Selection of the Experiment Population. The seventh grade was selected as the level on which to conduct this experiment simply because Mellon chose seventh graders. An important design feature of this experiment was that the experimental group was required to write out sentences virtually identical to those written out by Mellon's experimental group. The advantages were obvious. Should the experimental group not achieve the growth hypothesized for it, an interesting question would arise concerning Mellon's study. Mellon claimed that his experimental group's growth was the result of the combining practice. Although common sense suggests that Mellon was correct, it is nevertheless possible that it was some unique combination of transformational grammar and sentence combining that led to the increase in "syntactic fluency."

If the present study's experimental group were to achieve significantly more growth in syntactic maturity than the growth achieved by Mellon's experimental group, another equally absorbing question would arise. Both groups would have experienced similar amounts of "combining" practice which, Mellon claimed, made the difference. How to account for the differences? Perhaps the grammar studied by Mellon's group, because it was more difficult than Mellon imagined, acted as an inhibiting agent on the sentence-combining practice done by his students. Equally interesting problems would arise if the experimental groups in both studies achieved approximately equal increases in syntactic maturity or fluency. This whole question will, of course, be returned to when the results of the present study are examined in the last chapter.

No claim is being made for the unique suitability for seventh graders of the sentence-combining practice described in this study. Indeed, it is the present researcher's opinion that oral sentence-combining practice could begin in the second grade and, perhaps, in written form in grade four. While the number of kernels to be embedded, their vocabulary and comprehension levels, and the cognitive-syntactic maturity of the children would obviously be of paramount importance to the curriculum writer, the arguments already cited for the attractiveness of the present study's sentence-combining practice would appear to retain their validity. Grammar-free sentence-combining, capitalizing on syntactic abilities that students already possess and conducted in an acceptant atmosphere in which students are the final arbiters of acceptability, should prove successful in elementary as well as in secondary schools.

Schoolwide scheduling constraints dictated that the seventh grade consist of four classes containing respectively seventeen, eighteen, twenty-four, and twenty-four students and that two of these classes meet at the same time. Fortunately, the administration was able to accommodate a request for all the classes to meet during the first three periods of the day. Also, the experiment population remained fairly stable. During the year only two students left Florida High, one each from an experimental and control class, and one student entered a control class in January, halfway through the academic year. Naturally, these students' inputs were not included in the final tabulations. The experiment population thus consisted of a total of eighty-three students.

Schoolwide scheduling constraints also necessitated this researcher's teaching two of the four seventh grade classes. It would have been more desirable to have had a teacher other than the present researcher conduct these classes. But since this was simply not feasible, it was decided that Mr. James Barnes, English Department head at Florida High School, would take the larger of the experimental classes (24) and the smaller control class (18), while this researcher would take the smaller experimental class (17) and the larger control class (24). In this manner the teacher-treatment influence was controlled to some extent.

Control of Outside Language Experiences. There was no practical way to control for the language experiences of the subjects outside their English classes. However, conversations with their social studies and science teachers in particular made it clear that, as might be expected, the students as a whole were given roughly equivalent writing and discussion assignments both in school and at home. An important and highly structured part of the subjects' English course was an almost

overwhelming emphasis on the importance of free reading. Students were encouraged to read as much as possible at home and were given approximately one day per week of in-class time in which to read books they themselves had chosen. The number of books read by students was highly gratifying, but there was no evidence of an appreciable difference in the number read by experimental or control classes. Although stringent efforts were made to ensure that the subjects were given identical writing assignments in their composition and literature classes, there was no way to directly control their extracurricular writing experiences. However, random assignment of subjects to the respective treatments, for which a book of random numbers was used, presumably would control for such extraneous factors.

The Treatments. In the spring of 1969 the English Department at Florida High decided to concentrate heavily on reading instruction in the seventh grade and to spend about one-third of the year on the teaching of reading skills and free reading, which entailed allowing students to read a book of their choice in class for an hour at least once per week. In addition, it was decided that two short units in literature would also be presented, as would units in composition, dramatics, library skills, and language study.

When permission was later granted for the present study to be conducted, plans had to be made to accommodate both an experimental and a control treatment. Since there were excellent curricular reasons for retaining the "spring" plan and no design problems requiring that plan's alteration, it was decided that the control group would study the units already outlined. The experimental treatment would consist of shortened versions of each of the units mentioned, as well as the unit on sentence-combining practice. For example, in examining the concept of fiction, both groups read and discussed a number of short stories. While the control group worked with five short stories, the experimental group studied just three.

The control group did not study any kind of grammar because previous research, including Mellon's own study, suggested that the systematic teaching of formal grammar, as Neil Postman (1967) so aptly put it, "does very little or nothing or harm to students . . ." (p. 1162). One of the outstanding observations in the Mellon study (1969) was that the practice sentences studied by his control group in their study of formal grammar

> represented immature types which junior high school composition teachers rightly exhort their students to avoid, although the experimenter finds without exception that all widely used seventh grade texts are limited to these puerile sentence types . . . (p. 38). [These stu-

dents] experience and perhaps emulate sentences far below their attained level of syntactic fluency. (p. 39)

Therefore, because of what Mellon called "their manifest undesirability," these sentences and the grammar study that requires them were systematically excluded from the control treatment.

The literature units studied by the control group concentrated on literary "forms"—fiction, with a heavy emphasis on the short story; nonfiction, stressing biography; and poetry, with an accent on modern works. The texts for these units were the following:

Adventures for Readers Book I, by Elizabeth C. O'Daly and Egbert W. Nieman (New York: Harcourt Brace Jovanovich, 1958);
Vanguard, by Robert C. Pooley, Virginia Belle Lowers, Frances Magdanz, and Olive S. Niles (Glenview, Ill.: Scott, Foresman and Co., 1967);
Perspectives, by Robert C. Pooley, Alfred H. Gromman, Frances Magdanz, Elsie Katlejohn, and Olive S. Niles (Glenview, Ill.: Scott, Foresman and Co., 1963); and
Reflections on a Gift of Watermelon Pickle, and Other Modern Verse, by Stephen Dunning, Edward Lueders, and Hugh Smith (Glenview, Ill.: Scott, Foresman and Co., 1966).

The control group's dramatics unit consisted of individual and small group improvisations, and the selection, rehearsal, and presentation of student written and protessionally written plays of proven popularity with seventh graders. Everyone wrote a short play. The best play in each group was chosen by thc group who set about rewriting, polishing, and finally presenting it to the rest of the class. Although the literary and aesthetic quality of the student plays was, to say the least, uneven, the students obviously had fun.

The reading course began with a heavy, four-week concentrated dosage in September and continued sporadically throughout the year. Students worked on an individualized basis at their own speed on a large variety of materials. Materials were provided for every reading level from second grade through college. These consisted of:

SRA Reading Laboratory IIIa, by Don H. Parker (Chicago: Science Research Associates, 1964);
General RFU Reading for Understanding, by Thelma Gwinn Thurstone (Chicago: Science Research Associates, 1969);
The Macmillan Reading Spectrum (New York: The Macmillan Company, 1965);
Reading Skill Builders (Pleasantville, N.Y.: Reader's Digest Services, 1960);
The Literature Sampler: Secondary Edition, by Rita McLaughlin (Chicago: Learning Materials, 1962); and

An extensive classroom library of approximately one thousand paperback books covering grades 3 through 12.

The "free reading" dimension of the reading course was regarded as much more important than its skill-building counterpart.

Students were both required and encouraged to read as much as possible both in class and at home. A "points" system that this researcher had used for five or six years in previous schools was instituted. A certain number of points were given for each book read by the student who reported on a three-by-five card. The points were allocated according to the number of pages, the size of the print, and the reading ability of the student. Because an able student might be given only three points for a book that a less talented colleague had been given five points for, both students would have to work at close to capacity to satisfy the stipulated minimum requirement. Extra credit was of course given to students who surpassed this minimum. The physical proximity of the books, the provision of free *in-class* time for reading, and constant book sharing experiences on a formal and informal basis with large and small groups all combined to create a highly satisfactory unit, according to an unsigned class-wide evaluation of that unit conducted in June.

The control group's language study unit consisted of teacher-made study sheets and exercises on vocabulary study, dictionary skills, punctuation, capitalization, and usage. Spelling was not taught systematically but was attended to on an individual basis in the students' work in composition.

The control group's composition course was divided into two separate sections. The first consisted of "journal" writing. Students were required to write two pages per week as a minimum with a maximum of four pages. It had been our experience with the majority of seventh graders in previous years that they found writing a burdensome chore. The journal writing was designed first of all to get them to write anything at all. Writing is, among other things, a physical act, and, as with most physical acts, practice is a necessary step on the road toward competency. Students were encouraged to write about themselves, about their hopes and aspirations, their doubts, their frailties, their pet hates, their favorite singers, their parents, their friends: anything and everything that pertained to their lives. Worthwhile writing usually stems from sincerity and commitment and relevance. Their teachers ceased to be "English" teachers and, instead, tried to be sympathetic "listener-readers." Students wrote only on right hand pages so that their teachers could respond and react on the left hand pages. Handwriting had to be legible—barely. Spelling was largely

ignored along with punctuation and other mechanical considerations. The focus was unremittingly on content. Students were encouraged to complain if their teachers didn't respond in writing enough. Of course, all the communications in the journal were held in the strictest confidence. A student could fill his two pages and then, if he so desired, forbid his teacher to read them by labeling the first page "DO NOT READ" and by putting a line through both pages, bottom left to top right. Although very few students had recourse to the "DO NOT READ" command, they obviously liked the idea that they *could.* We felt this was an excellent writing unit; so did the students.

These same students were not so enchanted by the second half of the composition unit, in which prewriting was stressed. Ideas, organization, style, mechanics, and spelling were discussed and graded for. Students were given an opportunity to write narrations, descriptions, and expositions. They were unimpressed. "Why can't we write journals? *That* was fun."

The experimental group was exposed to all of the units described here. Their units were simply shorter. They worked on fewer exercises in their language study and read only one biography instead of the two read by their control counterparts. Their reading course was shorter, as were their literature units. In dramatics they had less time to work on their plays and presented only one play on stage. They were given less instruction in composition. However, they wrote exactly the same number of compositions as the control group, and they wrote an identical number of assignments in their literature study.

The experimental group worked on nineteen lessons which taught sentence-combining techniques and provided abundant practice in sentence combining. The text was called *Sentence Combining* and contained 111 pages of text and exercises. The students were directed, workbook fashion, to write all the required exercises directly on the pages of the text. The students used ring binders to keep the lessons distributed to them during the year. The ring binders were kept in the classroom so that they could be checked. When homework was assigned, the students took home only the relevant lesson sheets.

A deliberate attempt was made to keep the text as brief as possible. Explanations of particular sentence-combining techniques and illustrative examples seldom went beyond half a page. The students rarely needed help with any of the lessons. Working with the actual sentence-combining problems consistently removed any difficulties encountered by the students.

The sentence-combining treatment lasted an average of one hour and a quarter per week in class, and the students spent about half an

hour per week on related homework assignments. These weekly totals are averages because sentence combining was interspersed with whatever other units were being taught. Students who completed their sentence-combining assignments early were also encouraged to read their novels in class. The students obviously enjoyed free reading, and this helped to keep the sentence combining interesting by association.

The first part of the sentence-combining text gave students practice in writing out simple sentences by matching separated subjects and predicates. (It is important to remember that although these procedures are being described for the reader in grammatical terms, such terms were *never* used in the lessons.) Then the students were given practice with the addition of adverbial phrases to sentences. This was followed by a series of short lessons giving students practice in converting sentences to negatives, questions, and passives. Here are a few examples which instructed the students to use a variety of the combining signals they had learned. Where appropriate, the desired answer is written out as sentence B.

A. The rattler (HOW) slithered (WHERE), bit the sleeping baby (WHERE), and (HOW) disappeared.

Instructions: In the following exercise write out as many sentences as you can, using *all* of the information.

A. My car broke down.
My car broke down during the winter.
My car broke down every Monday morning.
My car broke down at five o'clock.

A. Some telephones are nearby. (THERE-INS + NEG + QUES)
B. Aren't there any telephones nearby?

A. Those dirty marks will fade away for some reason. (NEG + WHY QUES)
B. Why won't those dirty marks fade away?

The second part of the sentence-combining text required students to master single-embedding problems. For example:

A. Peter noticed SOMETHING.
There were nine golf balls in the river. (THAT)
B. Peter noticed that there were nine golf balls in the river.

Lesson Thirteen presented a particularly difficult problem for this researcher. What combining signal would, in non-grammatical terms, enable students to convert an adverb to an adjective and change its position in the transformed sentence? For example, how to get students to change "The child shivered violently" to either "The child's violent shivering . . ." or "The violent shivering of the child . . ." so that

either could be inserted in another sentence? The solution was so obvious that it eluded this researcher for a long time. Almost every adverb, when changed to an adjective, drops *ly*. Therefore, the parenthetical command (~~LY~~) was called "LY with the cross through it" in the lessons. The student followed the (~~LY~~) command and positioned the new word according to where it appeared in parentheses. For example,

A. The child shivered violently ('S + ~~LY~~ + ING)
B. The child's violent shivering. . . .

A. The child shivered violently (~~LY~~ + ING + OF)
B. The violent shivering of the child. . . .

The single-embedding problems were followed by multiple-embedding problems that required the students to transform and embed two, three, four, or more kernel sentences into a single sentence. For example:

A. SOMETHING should tell you SOMETHING.
John has not called in five days. (THE FACT THAT)
You are not going steady anymore. (THAT)
B. The fact that John has not called in five days should tell you that you are not going steady anymore.

Illustrative sentence-combining problems can be found in Appendix A.

The present researcher did not perform an actual count of the different forms of practice exercises because his exercises were virtually identical to those of Mellon, who performed that very laborious count. Mellon stated that the 904 kernel sentences used in the sentence-combining problems were presented in such a way that their proportions would be approximately equal to the proportions of transform types found in normal eighth grade writing. Mellon's 602 exercises consisted of 123 pretransformational basic sentences, 130 simple transformations, 68 separate complex transformations and 281 sentence-combining problems, with 98 single-embedding problems and 183 multiple-embedding problems. Generally speaking, then, the present study exposed its experimental students to a roughly equivalent number of problems.

A very important and perhaps crucial dimension of the experimental treatment was the nature of the classroom activities and the atmosphere in which the combining practice was conducted. Historically, usage and grammar drills have been negatively oriented, concentrating on errors instead of building confidence. Students learned to think in terms of "red ink" teacher comments. Especially at the beginning of this study, there was almost no concern at all with error. A student who perhaps had produced a good English sentence but not the one desired in the exercise was rewarded with an approving smile or nod. "That's a good

sentence, John" or a similar comment always accompanied the teacher's reaction to the exercise error. Most of the students needed a feeling of assurance, a sense of predictable success when they were faced with a sentence-manipulation problem.

In the first few lessons the teacher was the center of the activities. But not for long. Since students who depend exclusively on the teacher to decide whether a sentence is acceptable or not are unlikely to develop the confidence necessary for success in sentence combining, it was decided to structure the teacher out of the remaining lessons. The teacher deliberately sat down when the classes were going over their sentences. If a student read out a sentence to which any member of the class objected, the class as a whole decided by a show of hands whether they agreed or disagreed with the complaint. Students discussed the issues raised and students decided on a sentence's suitability. Only if there was no clear majority did the teacher step in with a hint or two. Indeed, if the majority decided that a particular sentence was acceptable and the teacher did not agree, it was decided to let the majority vote carry the day. One questionable sentence was of little importance when compared to the evident satisfaction the students derived from overruling their teacher. Producing sentences that were intellectually satisfying and grammatically correct most of the time gave the students the desired confidence and a positive attitude towards sentence production.

In addition to writing out every sentence, the students practiced choral readings of approximately one-third of the completed sentences. This was usually performed in a very relaxed atmosphere where a reasonable amount of student clowning was not frowned upon. Sometimes the exercises were gone over in small groups whose population was constantly changed. Often students volunteered to run the lesson and supervise any discussions. Generally speaking, a variety of techniques was consciously and very deliberately used to keep the exercises interesting. The length of the lessons ranged from ten to forty minutes. When the exciting dramatics unit began in January, the students seemed reluctant to get back to the combining exercises, claiming that they were too busy with their free reading, which was beginning to catch on then, and their play acting and writing. The teachers decided to postpone the sentence combining until the first week in February. This was a wise decision, for the students returned to it with a revived interest and enthusiasm that they never again lost.

Strenuous efforts were made to expose both the experimental and control classes in the experiment to the same kind and variety of classroom procedures. The experimenter discussed strategies with Mr.

Barnes on a systematic weekly basis, and both teachers visited each other's experimental and control classes periodically throughout the year.

Measurement

Ability. The students' ability was measured by the California Test of Mental Maturity (IQ scores, mean 100, SD 16) and by their score on words per T-unit calculated from the first ten T-units in each of the five pre-treatment compositions that they wrote.

Syntactic Maturity. In order to measure the syntactic maturity of the subjects' free writing, it was necessary to obtain a representative sample of that writing. Studies have shown that a writer's performance can vary because of day-to-day fluctuations and because of the mode of discourse. Kincaid (1953) discovered that, at least with college freshmen, the day-to-day writing performance of individuals varies, especially that of better writers. Anderson (1960) found that 71 percent of the fifty-five eighth grade students he examined on eight different occasions "showed evidence of composition fluctuation" and concluded that a writer variable must be taken into account when rating compositions for research purposes. Frogner (1933), Seegers (1933), and Hunt (1964) have shown that a writer's sentence structure is affected by the mode of discourse he is using—argumentation, exposition, narration, or description. Clearly, then, the topic and mode of discourse should be varied.

There have been no definitive studies done on ideal sample size. Chotlas (1944) discovered that 1000-word samples written by junior high school students were as reliable as 3000-word samples. Anderson (1937) showed that the 150-word samples used by LaBrant were unreliable and suggested samples several times larger. O'Donnell and Hunt (1970), used a 300-word sample for the writing of fourth graders. Using a method similar to O'Donnell and Hunt's, the present researcher sampled 10 percent of the pre-test compositions of his experiment population. Words per T-unit, the most sensitive measure of syntactic maturity in school children, was used to determine a reliable sample size. It was discovered that a sample just over 400 words in length was as reliable an indicator of average T-unit length as was a 1000-word sample. Since Hunt's eighth grade students wrote T-units approximately 11 words in length, it was decided to collect per-student samples fifty T-units in length of pre- and post-treatment writing. Hopefully this would result in samples approximately 500 words in length.

Since it was desirable that the students' compositions represent their

own writing ability, all of the compositions at pre- and post-tests were written in class under teacher supervision, thus eliminating potential help from parents or friends. Teachers distributed the printed topic sheets and read aloud the information contained in them while the students read them silently. All of the students were supplied with lined legal-size paper. No attempt was made to influence the students to adopt an unnatural writing style, nor were they told that their sentence structure would be singled out for analysis. Indeed, they were never told that this was an experiment. The topic sheets were headed "7th Grade, Composition Evaluation." In October the students were told that these compositions would be examined to help their teachers plan a composition program based on their particular needs. They were not told that they would be examined again in May. The post-test compositions were presented as an evaluation instrument to see how much they had improved their writing ability since the start of the school year. The students were encouraged to write rough drafts and to revise their sentences in any way they thought fit. The compositions were written during the first two weeks in October and the last two weeks in May.

The students in Mellon's experiment wrote nine pre- and post-test compositions. While a variety of modes of discourse was desirable, it was thought that this was an excessive amount of writing considering the fact that it was conducted in an environment where no composition instruction took place. Some students might have asked themselves how the teacher could grade nine compositions for each student right at the end of school. This researcher rather arbitrarily decided that five compositions were as many as the students would tolerate and therefore five topics were devised in consultation with Mr. Barnes. Following Mellon's methodology, each topic was represented in parallel A and B forms. Students who received one of these forms in October were given the other form in May. To avoid any systematic bias, half of one class were given the A form, the other half the B form. The process was reversed at the post-test so that half of both the experimental and control groups were writing on the same topics at any given time. The topics ranged over the three modes of discourse—narration, description, and exposition. Forms A and B of the five topics can be found in Appendix C exactly as they were given to the students.

The student writing was segmented and analyzed by the experimenter. The first ten T-units from each of a student's five compositions comprised the sample of fifty T-units per student per test.

The following are the rules used to segment each student's writing into T-units: one main clause plus any subordinate clause or nonclausal

structure that was attached to or embedded in it counted as one T-unit. Fragments which resulted from the omission of a word counted as a T-unit. The experimenter supplied the missing word. Other fragments were discarded. Unintelligible strings of words, referred to by Hunt (1965, p. 6) and O'Donnell, et al. (1967, p. 39) as "garbles," were discarded.

A very real difficulty arose when directly quoted discourse introduced by such an expression as "He said . . ." was encountered. Mellon discarded the "speaker tag." This experimenter was unhappy with such a procedure because it soon became difficult to define the expression "speaker tag." There is no great loss when an expression like "He said" is discarded. But what about the following example encountered in the analysis: "Clutching the knife tightly in his bleeding hand, Joe painfully crawled towards the opening and said, 'I surrender.'" Exactly what is the speaker tag here? Technically speaking, it would include every word from "Clutching" to "said." Surely this is not a two word T-unit with a sixteen-word speaker tag discarded!

Hunt stated that "there is some reason, then, to tabulate direct discourse along with noun clauses" (1965, p. 75). It would be easy to imagine directly quoted discourse consisting of a dozen sentences. Counting all of them as noun clauses would also be unsatisfactory. A compromise was reached by counting the first expression after "He said" as a direct object, because it seemed to satisfy the minimally terminable requirement for a T-unit. For example, the following discourse—

> Marsha said, "I really like you, John. However, Clarence's father is a millionaire and I like the idea of Palm Beach."

—would have been segmented into three T-units—between "John" and "However," and between "millionaire" and "and." The advantage of such a procedure was that it retained as much of the student's original writing as possible.

Mellon counted clauses of condition, concession, reason, and purpose as separate T-units because he believed that logical conjunctions behave much like coordinate conjunctions. In addition, he discarded clauses with repeating predicate phrases because he claimed they were elliptical and therefore vacuous. This experimenter remained unconvinced by Mellon's reasoning in either case and, therefore, retained Hunt and O'Donnell's simpler and more convincing methodology.

If a student failed to produce ten T-units in any composition, the shortage was made up by segmenting extra T-units written in his other compositions. Since the compositions were numbered from one

to five, the search for "extra" T-units always began with the first composition and proceeded until the shortages were eliminated. A few students failed to produce the required fifty T-units. It was not felt that this shortage compromised the adequacy of the sample because all of the computations were either converted to a base of one hundred T-units or expressed as a ratio of a certain number of words per T-unit or per clause.

In this experiment the six factors of syntactic maturity were calculated in the following manner:

Words per T-unit. This figure was obtained by dividing the number of words by the number of T-units. Compound nouns written as one word counted as one word. Compound nouns written as two words and hyphenated word pairs counted as two words. Phrasal proper names counted as one word. Dates like *June 21* or *July 2* counted as two words. Contractions such as *he'd* or *shouldn't* counted as two words.

Clauses per T-unit. This figure was obtained by dividing the number of subordinate and main clauses by the number of main clauses.

Words per clause. This figure was obtained by dividing the number of words by the number of subordinate and main clauses.

Noun clauses per 100 T-units, adverb clauses per 100 T-units, and adjective clauses per 100 T-units. These figures were obtained by dividing the total number of each type of clause by the number of T-units, quotient times 100.

The T-unit segmentation and the frequency counts were performed by this experimenter, who conducted a systematic series of spot checks and also rechecked every T-unit longer than fifteen words.

Writing Quality. An evaluation of the general quality of the compositions written by the experiment sample was clearly desirable. But several problems presented themselves. This researcher had neither the time nor the resources to arrange for the evaluation of all the compositions written. A further difficulty was the notorious unreliability of composition ratings. And there was also the problem of securing the services of a sufficient number of experienced evaluators for a satisfactory length of time to do the job properly.

With these problems very much in mind, it was decided that the system of forced choices between matched pairs of compositions would be utilized. Members of the control group were listed and numbered in ascending order of IQ for both boys and girls. A similar list was compiled for the experimental group. A subject was randomly chosen from the control group and a subject of the same sex and approximately equal IQ (within three or four IQ points) was chosen from the experimental group to make up a matched pair. Thirty matched pairs of sub-

jects, comprising sixty compositions in all, were secured in this fashion.

Although the students had written five pre- and post-treatment compositions, it was decided that only the post-treatment compositions would be evaluated since we were primarily interested in the two groups' post-treatment writing ability. The most typical writing taught in seventh grade and the modes of discourse which seventh graders are apparently happiest with are narration and description. Therefore, Composition 1, a narration, and Composition 2, a description, were chosen as the compositions to be evaluated. The matched pairs of subjects were then divided into high and low according to IQ and assigned in balanced form to achieve approximately equal numbers of the same sex and ability level, fifteen to each of the two compositions.

For the purposes of the present experiment the system of forced choices of matched pairs had several advantages over a rating scale. It enabled a direct one-to-one comparison to be made of the experimental and control group's writing, which was of course the major purpose of the evaluation. It had the further advantage of being very easy to administer. Also, the evaluator's task was made much simpler. There was no need for him to read a decent sample of all the compositions before deciding what precisely a 3, for example, meant in a 1-to-5 rating scale. He seldom needed to read a composition more than once. And he had only to decide which composition was, in his opinion, better than its partner. This could be accomplished rapidly and efficiently. Braddock (1963) warned that "fatigue may lead raters to become severe, lenient or erratic in their evaluations . . ." (p. 11). Since all eight evaluators easily completed the task within one hour, the fatigue question never arose. The evaluators were eight experienced English teachers who were attending Florida State University during the summer in 1970. All of the teachers volunteered for the project. Table 2 gives a brief idea of the length and type of their teaching experience. The average teaching experience of these five females and three males was just over six years. Of course, these evaluators had no knowledge of the nature of the present experiment. They were simply told to make a single judgment on the overall quality of the compositions in each pair, basing their decision on ideas, organization, style, vocabulary, and sentence structure.

Since the evaluators met during the same morning in one classroom, the experimenter was able to explain the procedures and answer any questions that arose, thus satisfying Braddock's stipulation that "It seems highly desirable to have all the raters working in the same or adjoining offices, where an investigator can be present and . . . insure that everything runs smoothly" (1963, p. 11).

Table 2
Academic Degrees Held or in Progress and Prior Experience of Teacher-Evaluators

Teacher	Sex	Degree Held	Degree in Progress	Grades Taught	Years of Teaching Experience
1	M	M.Ed.	Ph.D., Eng. Ed.	7–12	16
2	F	B.S.	M.A.	10	1
3	F	B.A.	M.S.	8–11	5
4	M	Advanced M.A.		9–12	5
5	F	B.A.	M.A.	10, 12	2½
6	F	B.A.	M.S.	11	7
7	M	B.Ed.	M.Ed.	10, 12	1
8	F	Advanced M.A.		1–3, 7–12	20

Both Stalnaker (1934) and Buxton (1958) claimed that rater training helps rater reliability. Buxton suggested that graders should review together a composition they have just rated to insure a common interpretation of their criteria. Braddock (1963) remarked on the frequency with which rater training is reported in studies which report high reliabilities (p. 14). Therefore, during an initial practice period the evaluators were given two matched pairs of compositions, one pair exemplifying very good seventh grade writing and the other, the contrary. On the blackboard, from left to right, were written *ideas, organization, style, vocabulary,* and *sentence structure.* After each item was discussed in turn, the evaluators were asked to choose the composition they preferred, basing their judgments equally on all five factors. They were to indicate their preference by making a large check at the top of the preferred composition. Then there ensued a discussion of the relative merits of each of the paired compositions to establish some sort of general agreement concerning the five criterion factors. These five factors were left on the blackboard, and the evaluators were encouraged to glance there occasionally to ensure that they were taking all five into consideration in their judgments.

This study was interested in the students' writing ability and not at all in their spelling, punctuation, or handwriting talents. In order to eliminate the possible effects of these extraneous factors on the evaluators' judgments, the thirty pairs of compositions were typewritten so that spelling and punctuation could be corrected. The corrections were made by a secretary at the University School. While fully aware that discourse can be punctuated in different ways that could possibly affect meaning, this researcher was satisfied that no bias was introduced

because all the punctuation and spelling changes were made by one person who was never aware of the group to which a particular composition belonged. In support of this position, Braddock stated that

> Even though raters are requested to consider in their evaluations such attributes as content and organization, they may permit their impressions of the grammar and mechanics of the compositions to create a halo effect which suffuses their general reactions. (A converse emphasis, of course, can just as easily create the halo.) (1963, p. 14)

A similar halo effect was reported both by Starring (1952) and by Diedrich, French, and Carlton (1961). Braddock, in a discussion of the factors that contribute to making a good composition, argued that

> However important accurate spelling may be in the clarity and social acceptability of composition, many of the factors of good spelling do not seem to be closely involved with the factors of good composition. (1963, pp. 49–50)

Thus there was ample justification both for typing the compositions and for eliminating any spelling and punctuation errors.

It was desirable to keep the evaluators in total ignorance of the group to which a particular composition belonged. This was achieved by using a complicated coding system that the evaluators could not possibly be expected to break. Although the composition pairs were each given a different number between one and thirty and were each stapled together, no two evaluators received their thirty pairs in the same order. Great care was taken to randomize the order, and the evaluators were instructed to ignore the order and simply to judge each pair according to the five criterion factors. In addition, the evaluators were instructed to put each stapled pair on the vacant desk beside them as soon as the preferred composition had been checked. This was designed to discourage the evaluators from imagining they had discovered a patterned sequence.

To enhance the reliability of their judgments, the evaluators were encouraged to read the compositions rapidly, according to the technique reported by Noyes (1963) for the College Entrance Examination Board.

CHAPTER 4

RESULTS

Assessment of Syntactic Maturity

Before the first hypothesis could be tested it was considered desirable to find answers to two questions: (1) Was there any evidence to indicate that the randomization procedures had not succeeded in equating the groups on the criterion measures? and (2) Had statistically significant growth occurred in the syntactic maturity scores of the control and experimental groups when analyzed separately?

To answer the first question, the pre-treatment mean scores on the six factors of syntactic maturity were compared by t-tests for two independent samples assuming unequal variances as described by Dixon and Massey (1969, p. 119). The results of these analyses, shown in Table 3, indicated that there were no significant differences between the groups, substantiating the assumption of equivalence of groups as a result of randomization.

Table 3

Comparison of Pre-treatment Mean Scores on the Six Factors of Syntactic Maturity: Experimental and Control Groups

	Experimental (N = 41)		Control (N = 42)			
Factors	**Mean**	**SD**	**Mean**	**SD**	**t-value**	**df**
Words/T-Unit	9.63	1.42	9.69	1.45	—.17(NS)	81
Clauses/T-Unit	1.36	.15	1.37	.15	—.42(NS)	81
Words/Clause	7.06	.69	7.05	.74	.06(NS)	81
Noun Clauses/ 100 T-Units	13.76	8.55	13.67	7.94	.05(NS)	80
Adverb Clauses/ 100 T-Units	14.34	6.37	14.24	7.95	.07(NS)	78
Adjective Clauses/ 100 T-Units	7.90	5.42	9.29	5.41	—1.16(NS)	81

t—test for two independent samples assuming unequal variances.
NS—not significant.

Table 4
Mean Pre-Post Change Scores on the Six Factors of Syntactic Maturity: Control Group

Factors	Pre	Post	Change	SD	t-value
Words/T-Unit	9.69	9.96	.27	1.27	1.37(NS)
Clauses/T-Unit	1.37	1.41	.04	.17	1.52(NS)
Words/Clause	7.05	7.03	—.02	.76	—.20(NS)
Noun Clauses/ 100 T-Units	13.67	15.85	2.18	9.57	.30(NS)
Adverb Clauses/ 100 T-Units	14.24	15.5	1.26	9.93	.83(NS)
Adjective Clauses/ 100 T-Units	9.29	10.16	.88	7.67	.74(NS)

t = 2.021, significant at the .05 level with 40 df.
NS—not significant.

Table 5
Mean Pre-Post Change Scores on the Six Factors of Syntactic Maturity: Experimental Group

Factors	Pre	Post	Change	SD	t-value
Words/T-Unit	9.63	15.75	6.12	2.50	15.68***
Clauses/T-Unit	1.36	1.84	.48	.28	11.07***
Words/Clause	7.06	8.55	1.49	.94	10.17***
Noun Clauses/ 100 T-Units	13.76	23.55	9.80	13.5	4.64***
Adverb Clauses/ 100 T-Units	14.34	29.01	14.67	12.3	7.66***
Adjective Clauses/ 100 T-Units	7.90	31.61	23.71	14.9	10.21***

***—significant at or beyond the .001 level.

To determine whether statistically significant growth had occurred in the control and experimental groups when examined separately, mean change scores, obtained by subtracting the pre- from the post-treatment mean scores, were analyzed by t-tests for correlated measures. The results of the analyses of the control group's pre–post change scores are shown in Table 4. Although five of the six factors of syntactic maturity showed evidence of increase, this growth was not statistically significant, substantiating Hunt's assertion that syntactic maturity develops with glacial slowness and is difficult to detect from one year to the next. The control group then would appear to have experienced normal growth in syntactic maturity.

The results of the analyses of the experimental group's mean pre–post change scores, shown in Table 5, indicated that highly significant growth had taken place on all six factors of syntactic maturity. The experimental group's mean pre–post change score of 6.12 words per T-unit was approximately five times the statistically significant increase reported for Mellon's experimental group. After the treatment Mellon's experimental group wrote on the average 11.25 words per T-unit, which was, according to the data shown in Table 1, typical of the writing of the average eighth grader in Hunt's study. The present study's experimental group wrote 15.75 words per T-unit, considerably more than the 14.4 words per T-unit reported by Hunt for the average twelfth grader.

There was therefore no evidence to indicate that randomization had not succeeded. In addition, only the experimental group showed statistically significant increases on the six factors of syntactic maturity.

The next step in the analysis of data was to test the first hypothesis, that the experimental group, which was exposed to the sentence-combining practice, would score significantly higher on the six factors of syntactic maturity than the control group, which was not exposed to the sentence-combining practice.

The post-treatment mean scores for the experimental and control groups on the six factors of syntactic maturity were compared by t-tests for two independent samples assuming unequal variances. Table 6 shows the results of these comparisons. It is evident from an examination of the t-values in Table 6 that the experimental group had established a highly significant superiority, at the .001 level of confidence, over the control group on all six factors. Since Hunt (1964, 1965) and O'Donnell, et al. (1967) have demonstrated that words per T-unit is the most reliable single index of syntactic maturity, it is interesting to note that words per T-unit yielded the highest t-value of all six of the factors.

Table 6
Comparison of Post-treatment Mean Scores on the Six Factors of Syntactic Maturity: Experimental and Control Groups

	Experimental (N = 41)		Control (N = 42)			
Factors	**Mean**	**SD**	**Mean**	**SD**	**t-value**	**df**
Words/T-Unit	15.75	3.00	9.96	1.64	10.88***	62
Clauses/T-Unit	1.84	.27	1.41	.16	8.72***	64
Words/Clause	8.55	1.02	7.03	.75	7.72***	73
Noun Clauses/ 100 T-Units	23.55	11.93	15.85	8.2	3.42***	71
Adverb Clauses/ 100 T-Units	29.01	11.15	15.5	7.67	6.41***	71
Adjective Clauses/ 100 T-Units	31.61	15.21	10.16	5.86	8.44***	51

t—test for two independent samples assuming unequal variances.
***—significant at or beyond the .001 level.

Table 7
Comparison by Grade Level of the Experimental Group's Post-treatment Scores on the Six Factors of Syntactic Maturity and Hunt's Normative Data

	Hunt's Normative Data		Post-treatment Experiment Group	
Factors	**Grade 8**	**Grade 12**	**Grade 7**	**Grade Level**
Words/T-Unit	11.5	14.4	15.75	12+
Clauses/T-Unit	1.42	1.68	1.84	12+
Words/Clause	8.1	8.6	8.55	12
Noun Clauses/ 100 T-Units	16.	29.	23.55	12—
Adverb Clauses/ 100 T-Units	16.	21.	29.01	12+
Adjective Clauses/ 100 T-Units	9.	16.	31.61	12+

Hunt (1965) reported that the average eighth grader wrote 11.5 words per T-unit, 1.42 clauses per T-unit, and 8.1 words per clause (p. 56). Hunt also reported that the average eighth grader wrote 16 noun, 16 adverb, and 9 adjective clauses per 100 T-units (1965, pp. 89–91). When these figures are compared with the post-treatment scores in Table 6, it is evident that the present study's experimental group wrote well beyond the syntactic maturity level typical of eighth graders.

To obtain some idea of the magnitude of the experimental group's growth on the six factors of syntactic maturity, the scores were compared, in Table 7, with the normative data reported by Hunt (1965) for eighth and twelfth graders. Table 7 also indicates by means of plus and minus signs the approximate grade level for syntactic maturity achieved by these seventh graders. Not only were the experimental group's mean post-treatment scores significantly greater than those of the control group, they were also distinctly greater than the norms reported by Hunt for eighth graders and at least similar to, and on four occasions superior to, Hunt's norms for twelfth graders. On only one factor, noun clauses per 100 T-units, were these seventh graders' average scores below those reported for twelfth graders.

Table 8 presents further evidence to support the assertion that the experimental group achieved significantly greater growth in syntactic maturity than that achieved by the control group. This table compared the pre–post change scores of the experimental and control groups by t-tests for two independent samples assuming unequal variances. Comparison of the change scores for words per T-unit yielded the highest t-value. Comparison with the change scores reported by Mellon (1969, p. 52) should serve as a useful indicator of the extent of this growth. Mellon reported that in mean words per T-unit his control group had increased by .26 and his experimental group by 1.27. The change in words per T-unit of 6.12 reported in Table 8 for the experimental group was over twenty times greater than that achieved by the present study's and Mellon's control groups, and approximately five times greater than the change reported for Mellon's experimental group.

Given the evidence cited above and the design of this study, it was concluded that the experimental group achieved significantly more growth in syntactic maturity than did the control group.

Several secondary questions, however, remained unanswered. It was considered desirable to determine whether the magnitude of the treatment effect could be related to the student's sex or ability level or to the influence of a particular teacher.

Since an important feature of the sentence-combining treatment was a deliberate attempt to structure the teacher out of the lessons as much

Table 8
Comparison of Mean Pre-Post Difference Scores on the Six Factors of Syntactic Maturity: Experimental and Control Groups

	Experimental (N=41)		Control (N=42)			
Factors	**Mean**	**SD**	**Mean**	**SD**	**t-value**	**df**
Words/T-Unit	6.12	2.50	.27	1.27	13.40***	59
Clauses/T-Unit	.48	.28	.04	.17	8.69***	66
Words/Clause	1.49	.94	—.02	.76	8.05***	77
Noun Clauses/ 100 T-Units	9.80	13.52	2.18	9.57	2.96**	72
Adverb Clauses/ 100 T-Units	14.67	12.26	1.26	9.93	5.47***	77
Adjective Clauses/ 100 T-Units	23.71	14.87	.88	7.67	8.76***	60

t—test for two independent samples assuming unequal variances.
**—significant at or beyond the .01 level.
***—significant at or beyond the .001 level.

Table 9
Comparison of the Mean Post-treatment Scores by Teachers on the Six Factors of Syntactic Maturity: Control Group

	Teacher 1 (N=18)		Teacher 2 (N=24)			
Factors	**Mean**	**SD**	**Mean**	**SD**	**t-value**	**df**
Words/T-Unit	10.00	1.34	9.92	1.86	.17(NS)	40
Clauses/T-Unit	1.40	.14	1.42	.17	—.48(NS)	40
Words/Clause	7.14	.67	6.95	.80	.86(NS)	39
Noun Clauses/ 100 T-Units	16.31	8.55	15.5	8.09	.31(NS)	36
Adverb Clauses/ 100 T-Units	14.62	6.64	16.17	8.44	.67(NS)	40
Adjective Clauses/ 100 T-Units	9.15	5.31	10.92	6.24	.99(NS)	39

t—test for two independent samples assuming unequal variances.
Teacher 1—Barnes. Teacher 2—O'Hare.
NS—not significant.

as possible, it seemed unlikely that any comparison of the influence of the two teachers would prove to be significant. Tables 9 and 10 confirmed this. Table 9 compared the control group's mean post-treatment scores by teacher on the six factors of syntactic maturity. Table 10 did the same for the experimental group. Analyses of the post-treatment mean scores indicated no significant differences between teachers.

Table 10
Comparison of the Mean Post-Treatment Scores by Teachers on the Six Factors of Syntactic Maturity: Experimental Group

	Teacher 1 (N=24)		Teacher 2 (N=17)			
Factors	**Mean**	**SD**	**Mean**	**SD**	**t-value**	**df**
Words/T-Unit	15.85	2.60	15.61	3.57	.24(NS)	28
Clauses/T-Unit	1.87	.23	1.8	.33	.82(NS)	27
Words/Clause	8.46	.91	8.67	1.18	—.61(NS)	29
Noun Clauses/ 100 T-Units	22.24	9.21	25.41	15.08	—.77(NS)	24
Adverb Clauses/ 100 T-Units	30.03	9.45	27.56	13.37	.66(NS)	27
Adjective Clauses/ 100 T-Units	35.04	15.71	26.76	13.47	1.81(NS)	37

t—test for two independent samples assuming unequal variances.
Teacher 1—Barnes.
Teacher 2—O'Hare.
NS—not significant.

While teacher influence had been predictably insignificant, it was not at all clear whether the treatment effect might depend on the sex of the students. Table 11 revealed no significant differences on control post-treatment scores between males and females, and Table 12 revealed the same result for the experimental group.

In Table 13 two aspects of the experimental treatment effect were considered: (1) Was there a relation between the experimental treatment effect as measured by pre–post change scores and IQ? (2) What was the correlation between post- and pre-test, post-test and IQ scores, and the combination of pre-test score and IQ versus post-test score?

1. Table 13 indicated a significant positive correlation, at the .01 level of significance, between pre–post change scores and IQ for words

Table 11
Comparison of the Mean Post-treatment Scores by Sex on the Six Factors of Syntactic Maturity: Control Group

	Male (N=21)		Female (N=21)			
Factors	**Mean**	**SD**	**Mean**	**SD**	**t-value**	**df**
Words/T-Unit	9.83	1.27	10.08	1.97	—.48(NS)	34
Clauses/T-Unit	1.39	.13	1.44	.18	—.97(NS)	37
Words/Clause	7.08	.67	6.98	.83	.41(NS)	38
Noun Clauses/ 100 T-Units	13.98	7.21	17.71	8.86	—1.5(NS)	38
Adverb Clauses/ 100 T-Units	15.29	7.41	15.71	8.11	—.18(NS)	40
Adjective Clauses/ 100 T-Units	9.75	5.21	10.57	6.55	—.45(NS)	38

t—test for two independent samples assuming unequal variances.
NS—not significant.

Table 12
Comparison of the Mean Post-treatment Scores by Sex on the Six Factors of Syntactic Maturity: Experimental Group

	Male (N=22)		Female (N=19)			
Factors	**Mean**	**SD**	**Mean**	**SD**	**t-value**	**df**
Words/T-Unit	15.02	2.97	16.60	2.89	—1.72(NS)	38
Clauses/T-Unit	1.79	.30	1.90	.23	—1.26(NS)	38
Words/Clause	8.38	.80	8.75	1.22	—1.13(NS)	30
Noun Clauses/ 100 T-Units	21.43	12.11	26.01	11.53	—1.24(NS)	39
Adverb Clauses/ 100 T-Units	27.71	12.26	30.51	9.82	—.81(NS)	39
Adjective Clauses/ 100 T-Units	30.16	17.90	33.29	11.62	—.67(NS)	36

t—test for two independent samples assuming unequal variances.
NS—not significant.

per T-unit. The correlation was significant at the .05 level for clauses per T-unit and adverb clauses per 100 T-units. This indicated that pupils with a high IQ tended to have a larger pre–post change score than students with a low IQ on the three variables mentioned.

Table 13

Correlations Between Pre- and Post-test Scores and IQ on the Six Factors of Syntactic Maturity for the Experimental Group (N=41)

	Correlations Between			
	Post-test and			Pre-Post
Factors	Pre-test	Pre-test and IQ†	IQ	Change and IQ
Words/T-Unit	.562**	.707**	.595**	.522**
Clauses/T-Unit	.227	.458**	.435**	.325*
Words/Clause	.459**	.543**	.411**	.234
Noun Clauses/ 100 T-Units	.160	.278	.251	.112
Adverb Clauses/ 100 T-Units	.103	.359	.343*	.318*
Adjective Clauses/ 100 T-Units	.240	.369	.333*	.245

†—this is a multiple correlation.
*—significant at the .05 level.
**—significant at the .01 level.

2. Table 13 also presented the correlations between the post-test, pre-test, and IQ scores, and the multiple correlation, R, between the post-test scores and the pre-test and IQ scores. The multiple correlation was significant at the .01 level for the first three variables. The correlations for the fourth, fifth, and sixth variables tended to be not significant, indicating large fluctuations in these variables. Therefore the post-test scores on words per T-unit, clauses per T-unit, and words per clause can be predicted from the pre-test score and IQ with multiple correlations of .71, .46, and .54 respectively.

Table 14 can be used to predict post-test scores from pre-test scores and IQ. The standard errors in the table indicate the accuracy of any single prediction.

Table 14
Multiple Regression of Post-test Score on IQ and Pre-test Score on the Six Factors of Syntactic Maturity: Experimental Group (N=41)

Factors	A	B_1	B_2	s	R
Words/T-Unit	—4.15	.104 (.028)	.862 (.259)	2.18	.707**
Clauses/T-Unit	.52	.0084 (.003)	.279 (.277)	.25	.458**
Words/Clause	2.06	.023 (.011)	.547 (.210)	.88	.543**
Noun Clauses/ 100 T-Units	—2.04	.208 (.143)	.168 (.221)	11.8	.278
Adverb Clauses/ 100 T-Units	—6.19	.291 (.128)	.187 (.265)	10.7	.359
Adjective Clauses/ 100 T-Units	—9.39	.334 (.180)	.460 (.439)	14.5	.369

$Y = A + B_1X_1 + B_2X_2$ where A = intercept
Y = predicted post-test score
X_1 = IQ
X_2 = pre-test score
B_1 = regression coefficient for IQ
B_2 = regression coefficient for pre-test score.
Numbers in parentheses are standard errors of regression coefficients.
R—multiple correlation coefficient.
s—standard errors of estimate.
**—significant at the .01 level.

Assessment of Writing Quality

The next step in the analysis of data was to test the second hypothesis: that the experimental group's compositions would be judged by eight experienced English teachers as significantly superior in overall quality to the compositions written by the control group.

Experimental Versus Control Compositions. To test this hypothesis, fifteen narrative and fifteen descriptive compositions were selected from both the control and experimental groups and paired by sex and level of IQ as described in greater detail in Chapter Three. Eight experienced English teachers were each given the thirty composition

pairs and instructed to indicate which composition in a pair was better in overall quality, according to the five criterion factors.

Assuming no difference between the paired compositions, the probability was .5 that a teacher would pick a composition written by one of the experimental group. Thus $P_i - .5$ indicated the difference between the observed proportion of experimental compositions picked by teacher *i* and the expected proportion. These proportions were tested by the χ^2 test. The calculations are summarized in Table 15.

The χ^2 for Teacher 1 in Table 15 was calculated in the following manner:

$$\chi^2 = \frac{2(O_1 - E_1)^2}{E_1} = \frac{2(22 - 15)^2}{15} = 6.53$$

The total χ^2 was equal to 42.00 with 8 degrees of freedom and was significant at the .001 level. This implied that the proportion $P = 169/240 = 0.7042$ of experimental compositions selected differed significantly from what would have been expected by chance, that is, $P = 0.5$. Therefore, the experimental group can be said to have written compositions which were judged to be significantly better in overall quality than those written by the control group. Thus, the second major hypothesis was confirmed.

Table 15

Comparison Between Number of Experimental Compositions Selected and Expected

Teacher	Number of Experimental Compositions		$\chi^2 = \frac{2(O_i - E_i)^2}{E_i}$
	Selected (O_i)	Expected (E_i)	
1	22	15	6.53
2	20	15	3.33
3	24	15	10.80
4	21	15	4.80
5	21	15	4.80
6	19	15	2.13
7	21	15	4.80
8	21	15	4.80
Total	169	120	42.00

Two other questions were of interest: (1) Did the teacher-evaluators judge the narrative and descriptive compositions differently with respect to the experimental treatment? and (2) Did these teacher-evaluators as a group agree in their rating of these compositions?

Narrative Versus Descriptive Compositions. Tables 16 and 17 illustrate the actual choices made by the eight experienced teacher-evaluators on the thirty pairs of compositions that had been matched by sex and IQ. Table 16 shows the choices made on the 15 pairs of narration compositions. Table 17 shows the choices made on the 15 pairs of description compositions.

Table 16

Experimental or Control Compositions Chosen by the Eight Experienced Teachers from Fifteen Matched Pairs of Narration Compositions

Teacher-	Composition Pair No.														
Evaluator	1	2	3	4	5	6	7	8	9	10	11	12	13	14	15
1	X	X	O	X	O	X	X	X	X	X	X	O	X	O	X
2	X	X	O	X	O	X	X	X	X	X	X	O	X	O	X
3	X	X	O	X	O	X	X	X	X	X	X	X	X	O	X
4	X	X	O	X	O	X	X	O	X	O	X	O	X	O	X
5	X	X	O	X	O	X	X	X	X	O	X	O	X	O	X
6	O	X	O	X	O	X	X	X	X	X	X	O	X	X	X
7	X	X	O	X	O	X	X	X	X	X	X	O	X	O	X
8	X	O	X	X	O	X	X	O	X	X	X	O	X	O	X

X indicates that the teacher preferred the composition written by a member of the experimental group.
O indicates that the teacher preferred the composition written by a member of the control group.

Table 17

Experimental or Control Compositions Chosen by the Eight Experienced Teachers from Fifteen Matched Pairs of Description Compositions

Teacher-	Composition Pair No.														
Evaluator	1	2	3	4	5	6	7	8	9	10	11	12	13	14	15
1	X	O	X	O	X	X	X	X	X	X	X	O	X	O	X
2	X	O	O	O	X	X	O	X	X	X	X	O	X	O	X
3	X	O	X	O	X	X	X	X	X	X	X	O	X	O	X
4	X	X	X	O	X	X	X	O	X	X	X	X	X	O	X
5	X	O	X	O	X	X	X	X	X	X	X	O	X	O	X
6	O	X	X	O	X	X	O	X	O	O	X	O	X	O	X
7	O	O	X	O	X	X	X	X	X	X	X	O	X	O	X
8	O	X	X	O	X	X	X	O	X	X	X	X	X	O	X

X indicates that the teacher preferred the composition written by a member of the experimental group.
O indicates that the teacher preferred the composition written by a member of the control group.

Calculations similar to those done in the comparison between experimental and control compositions were done for the narrative and descriptive compositions and led to a $\chi^2 = 25.07$ for the narrative compositions and a $\chi^2 = 20.80$ for the descriptive compositions, both of which were significant at the .01 level. Therefore, the narrative experimental compositions were significantly better than the narrative control compositions, and, similarly, the descriptive experimental compositions were significantly better than the descriptive control compositions.

The second aspect of the comparison between the narrative and descriptive compositions was whether the eight teachers selected the same proportion of experimental compositions in the narrative group as in the descriptive group. These results were summarized as follows:

	Essay Selected			
	Experimental	**Control**	**Total**	**Proportion**
Narrative	86	34	120	.7167
Descriptive	83	37	120	.6917
Total	169	71	240	
Proportion	.7042	.2958		

The proportions, .7167 for the narrative compositions and .6917 for the descriptive compositions, were compared by the χ^2 test for a 2×2 contingency table. The result was $\chi^2 = .1800$, which was not significant. Therefore, the proportion of experimental compositions selected did not differ significantly in the narrative and descriptive groups.

Agreement of Teachers. It was of interest to see how well the teachers agreed in their assessment of the composition pairs. Since there was no explicit reference to a measurement of this type except Cochran (1950), Gerald vanBelle of the Department of Statistics at Florida State University developed the following measure: Consider the proportion of times (P_j) the teachers selected the experimental composition in the *j*th pair. There would be perfect agreement if $P_j = 1$ or $P_j = 0$. There would be maximum disagreement if $P_j = 0.5$. Therefore, a reasonable statistic to test agreement is

$$\chi^2 = 2\frac{\sum_{j=1}^{N} (mP_j - m/2)^2}{m/2}$$

where m = number of teachers
and N = number of composition pairs.

Under the null hypothesis of no agreement between teachers, this quantity has approximately a χ^2 distribution with N degrees of freedom.

For the data in this study $\chi^2 = 157.5$ with 30 degrees of freedom. Since this result was significant at the .001 level, the hypothesis of no agreement between teachers was rejected. Therefore there was substantial agreement among the eight teachers who judged the overall quality of the compositions.

Summary

Analysis of the data on the six factors of syntactic maturity indicated the following:

1. There was no evidence to indicate that the randomization procedures had not succeeded.
2. The experimental group had experienced highly significant growth, at the .001 level, on all six factors of syntactic maturity.
3. The experimental group established a highly significant superiority, at the .001 level, over the control group on all six factors.
4. The experimental group wrote well beyond the syntactic maturity level typical of eighth graders and, on five of the six factors of syntactic maturity, their scores were similar to those of twelfth graders.
5. The treatment effect could not be related to the influence of a particular teacher or to whether a student was male or female.
6. Although students with a low IQ achieved highly significant increases in syntactic maturity, those with a high IQ tended to do even better.

Analysis of the data on the overall quality of the writing sample as judged by the eight experienced English teachers indicated the following:

1. The experimental group wrote compositions that were judged to be significantly better, at the .001 level, in overall quality than those written by the control group.
2. Both the narrative and descriptive compositions were significantly better, at the .01 level, than their control counterparts.
3. The proportion of experimental compositions selected did not differ significantly in the narrative and descriptive groups.
4. There was substantial agreement between the eight teachers who judged the overall quality of the compositions.

CHAPTER 5

CONCLUSIONS AND IMPLICATIONS

The present study was designed to measure the effect of written and oral sentence-combining exercises on the free writing of a seventh grade experimental group. The experimental group was given intensive practice in combining groups of kernel statements, by addition and deletion, into single sentences which were structurally more complex than those students would normally be expected to write. In order to facilitate the sentence-combining operations a series of signals capitalizing on the students' inherent sense of grammaticality was developed. An important, perhaps crucial, dimension of these signals was that they were in no way dependent on the students' formal knowledge of a grammar, traditional or transformational. Also important was an acceptant classroom atmosphere designed to allay possible syntactic fears and to produce a student confident in his ability to manipulate sentence structure. Specifically, the present study was designed to answer two questions. In comparison with the control group who were not exposed to the sentence-combining exercises, would the experimental group in their free writing (1) write compositions that could be described as syntactically more elaborated or mature? and (2) write compositions that would be judged by eight experienced English teachers as better in overall quality?

Conclusions

As a result of the analyses of data presented in Chapter Four, it was concluded that the experimental group wrote compositions which were syntactically different from the compositions written by the control group. The experimental group wrote significantly more clauses and these clauses proved to be significantly longer. As a consequence the experimental group wrote T-units which were significantly longer than those of the control group. When compared with the normative data presented by Hunt (1965), the experimental group's compositions showed evidence of a level of syntactic maturity well beyond that typical of eighth graders and in many respects quite similar to that of twelfth graders.

When eight experienced English teachers were asked to judge the overall writing quality of thirty pairs of experimental and control compositions, sixty compositions in all, that had been matched by sex and IQ, they chose a significantly greater number of the experimental compositions. Therefore, it was concluded that the experimental group wrote compositions that were significantly better in overall quality than the control group's compositions.

Given the design features of the present study, it seems reasonable to attribute the superior performance of the experimental group to the experimental treatment. For these reasons it has been judged that sentence-combining practice that is in no way dependent on formal knowledge of a grammar has a favorable effect on the writing of seventh graders.

Implications

The present study has demonstrated that the writing behavior of seventh graders can be changed by certain written and oral language experiences and that it can be changed fairly rapidly and with relative ease. In a sense this assertion questions the belief that growth in writing ability is *necessarily* a slow and difficult process. In showing that significant qualitative and syntactic gains can be achieved in approximately eight months, the present study suggests that, at least for seventh graders, a part of the composing process is directly amenable to alteration.

In the Epilogue to his NCTE report written two years after his original study, Mellon repeatedly asserted that "the sentence-combining practice had nothing to do with the teaching of writing" (1969, p. 79). The present researcher rejects such an assertion. Both Mellon's and the present study's experimental groups practiced writing sentences. The sentence-building process involved semantic as well as syntactic considerations: How does it sound? Does it make sense? Does it include all the input information (the kernels)? All of these questions, which surely include rhetorical considerations too, were an integral part of both treatments. At least by implication, both treatments favored sentences that were syntactically more mature than those the students were accustomed to producing. Football coaches have their players practice play after play in an "a-game" setting, often with no opposition, so that they will be able to execute efficiently in an actual game. Surely the coach at practice is teaching football. Similarly, students exposed to sentence-building exercises, even in an "a-rhetorical" setting, are in a very real sense being taught writing. Both treatment groups ended up

writing sentences that were syntactically more mature. The present study's experimental group wrote compositions that were judged better in overall quality. The acceptance or rejection of Mellon's overall hypothesis depended entirely on whether his students wrote syntactically more maturely. Mellon reported that the students were generally able to complete the sentence-combining exercises. But the crucial question was whether they had developed syntactic manipulating skills that would show in their writing. Mellon's study was clearly concerned with the teaching of writing skills. It is, therefore, difficult to understand how sentence-building exercises can be defined out of the teaching of writing.

Indeed, sentence combining has both theoretical and practical attractiveness when considered as part of a composition program. Rhetoric and sentence-combining practice should be viewed not as mutually exclusive or even discrete but rather as complementary. Gleason (1962), in an article discussing the place of language study in the curriculum, argued that the choppy style and the run-on style

> are basically the same. Each chooses one device to the exclusion of all others. The style is bad, not because of any individual choice, but because of the monotonous patterning. . . . to produce a good style it would be necessary to select out of a wider stock of available devices, and to work them all into an appropriate, pleasing over-all pattern. (p. 5)

Gleason went on to ask what a student must be made aware of if he is to understand and control style.

> He must know the options. The wider his repertoire and the deeper his understanding of the peculiarities of each, the better equipped he is to write. . . . As in teaching a foreign language, the accurate, casual control of patterns comes out of specific patterned drill and conscious manipulations. (pp. 5–6)

This is precisely what sentence combining provides. It expands the practical choices, the options truly available to the inexperienced young writer *when he needs them.* Christensen (1968*a*) claimed that "Grammar maps out the possible, rhetoric narrows down the possible to the desirable or effective" (p. 572). Sentence combining helps the writer enlarge the "practical-possible" so that it can be utilized during the composing process. The young writer, who has been exposed to sentence-building practice and who is developing into what was earlier called "the student as syntactic authority" as a result of intensive experiences with the manipulation of sentence structure, should be in a better position to deal with run-on or choppy styles. Armed with an

expanded practical repertoire of syntactic choices, he would be better able to avoid "monotonous patterning" and to work his "wider stock of available devices" into "an appropriate, pleasing overall pattern" as advocated by Gleason. Clearly a desirable curricular outcome for the teacher of writing.

Although the findings of the present study relate specifically to seventh graders, there is no obvious reason for assuming that sentence-combining practice should not be used in elementary and senior high school, as well as in junior high school.

The English department at Florida High School spent a good deal of time planning the seventh grade language arts program for the control group. (Remember that the experimental group was exposed to shortened versions of each of the control group's units.) And yet, despite the sophistication of the control group's program, with its small classes, well-qualified, experienced teachers, an abundance of free reading, carefully planned instruction in composition, and a relaxed atmosphere in which student talk and classroom interaction were encouraged, the control group showed only "normal" growth—.27 words per T-unit—in syntactic maturity, very similar to Mellon's control group which increased by .26 words per T-unit. If the control group's program had such a negligible effect on their syntactic maturity and overall writing quality when compared to the experiences of the experimental group, it seems reasonable to advocate the use of sentence-combining practice with, at the very least, seventh graders. The case for the efficacy of sentence-combining practice becomes even more attractive when the results of research in composition are reviewed. Neither Braddock (1963) nor Meckel (1963) uncovered a single study reporting a statistically significant composition treatment effect. Since the present study did discover a significant composition treatment effect, its sentence-combining system, which enables students to build sentences and manipulate syntax with greater facility, should surely be utilized in our schools.

In elementary school, simple adjective and relative clause insertions and repeated subject and verb deletions could be practiced orally in, perhaps, second grade. Written exercises could start in third or fourth grade. The present study's sentence-combining signal system can easily be expanded to incorporate a wider range of syntactic structures which could be practiced in junior and senior high school.

Students exposed to sentence-building techniques could use these syntactic manipulative skills at the prewriting or rewriting stage in their work in composition. They would be better able to "unchop" the choppy sentence and eliminate the run-on sentence. One can readily

envisage individual or classwide work on improving sentences or even paragraphs in a rhetorically oriented setting. Students could practice rewriting whole paragraphs, given either in kernel form or in a choppy or overly elaborate style. Experienced in sentence manipulation and trained to think in rhetorical terms, they would be in a better position to make meaningful rhetorical choices because they would have a wider repertoire of syntactic alternatives from which to choose.

The present researcher certainly agrees with Mellon's statement that sentence-combining exercises could be regarded as "a valuable addition to the arsenal of language-developing activities Moffett (1968) includes in his language arts program" (1969, p. 80). Whether these activities are "naturalistic" or "non-naturalistic" is, perhaps, irrelevant. The crucial questions are (1) Would they work? and (2) Would students enjoy them? A skillful teacher should be able to ensure that both questions are answered in the affirmative.

Practice with intensive sentence-manipulation exercises need not be restricted to the lower grades. Hunt's data, shown in Table 1 (p. 22), indicate a wide gap between the syntactic maturity level of twelfth graders and that of superior adults. Indeed, *The Christensen Rhetoric Program* (1968*b*), although heavily dependent on the students' prior knowledge of grammatical terminology, does teach sentence-building operations in order to improve college freshmen's writing ability.

Although Christensen agreed with Mellon that a mature style can be taught, he strenuously disagreed with what he called Mellon's conception of good style. He criticized Mellon for concentrating on relativization and, especially, nominalization, and also suggested that "we shouldn't teach subordination as it is hard to read" (1968*a*, p. 576). Christensen based this argument on an examination he made of modern professional and semiprofessional writers. These writers, Christensen claimed, wrote what he called "cumulative sentences," which feature a high proportion of final free modifiers and are indicative of a mature style. However, another researcher, Johnson (1969), after analyzing the prose of a different group of professional writers—a very prestigious collection indeed—and comparing them with Christensen's "best" writer, Halberstam, concluded that

> If we are to measure the degree of skill in a writer by the percentage of words he has in free modification, we should rate Cather, Fitzgerald, Forster, Isherwood, Baldwin, Auden and Orwell as less skillful than Halberstam. (p. 163)

Johnson also suggested that "students had best devote far more time to mastering subordination than Christensen would have them do" (p.

161). She noted that Edmund Wilson's style is one that is not cumulative but periodic and that Wilson

> depends for modifications, not on verbal clauses, appositives and absolutes so much as on relative and subordinate clauses. . . . (p. 162)

It is obvious from the evidence advanced, both by Christensen and Johnson, that we are a long way from defining satisfactorily a mature style or styles. Relativization and nominalization, final, medial, and initial free modifiers, short base clauses, all would appear to have their place in any definition of what constitutes a mature style.

What is bad about any style is its obviousness. Repeated cumulative sentences draw attention to themselves; their lack of variety only has unfortunate stylistic consequences. Therefore it would surely be a mistake to favor any one particular syntactic pattern to the exclusion of other possible patterns. Syntactic manipulative exercises should exploit the entire range of syntactic alternatives allowed by the grammar of English. What the young writer needs is as much practice as possible on every conceivable combination of syntactic operations.

In *Notes Toward a New Rhetoric* (1967) Christensen raised an interesting point that may help to explain something that the present researcher noticed in an entirely subjective examination of the post-treatment compositions. Christensen claimed that "solving the problem of *how to say* helps solve the problem of *what to say* . . ." (p. 5). Does this mean that form can, in some sense, generate content? It was evident to this researcher that the post-treatment compositions written by the experimental group had much more detail, more "meat" to them. The treatment group seemed to "see" more clearly. They had more to say. Perhaps the syntactic manipulative skill the students had developed, because it entailed a wider practical set of syntactic alternatives, *invited* or *attracted* detail. Perhaps knowing *how* does help to create *what*.

An alternative explanation seems plausible. Since the experimental group had become more skillful manipulators of syntax, perhaps their fear of syntax had dissipated. Confidence is very likely a self-generating process, feeding on itself. Released from syntactic roadblocks, confident, seeing a wider range of choices, the student's mind could grapple, at ease, with additional syntactic-semantic considerations. It is of interest to note that although the sentence-combining exercises did not include practice with adverb clauses, the experimental group produced a significantly greater number of adverb clauses in their free writing. The "confidence" factor has a theoretical attractiveness that invites further study. An important dimension of the present study was

the systematic attempt to build student confidence by accentuating the positive. Perhaps grammar study and too much concern with error build barriers between the beginning writer and the composing process. Sentence combining concentrates on student success. It not only has students write, it shows them *how*.

Since this researcher is advocating work with cumulative sentences as well as with sentences similar to those in the present study, it might be of interest to illustrate how readily adaptable the present study's sentence-combining signals are to Christensen's system or similar programs. In an article describing and evaluating the latest developments in rhetorical theory, McCrimmon (1968, p. 128) cited his favorite example of a cumulative sentence, written by one of Christensen's students. Reduced close to basic kernel form, with sentence-combining signals added, it would look like this:

> A girl develops from the sheet of walk.
> The girl is *small*.
> The girl is a *Negro*.
> The walk is *glare-frosted*.
> The girl is *walking barefooted*. (,)
> *Her bare legs* are *striking* the cement. (,)
> Her legs are *recoiling from the cement*. (AND)
> The cement is *hot*.
> *Her feet* are *curling in*. (,)
> *Only the outer edges* of her feet are
> *touching* the hot cement. (,)

Note that the only additional signals that had to be developed are (AND) and (,). The (AND) simply means insert an *and* where appropriate on that line. Perhaps the comma signal (,) is not really necessary. Remember that underlined (italicized) words are retained and the remainder of the sentence deleted. The final sentence is rather easy to produce:

> A small Negro girl develops from the sheet of glare-frosted walk, walking barefooted, her bare legs striking and recoiling from the hot cement, her feet curling in, only the outer edges touching.

Similarly, the following example from Sinclair Lewis can be readily handled by the development of two additional signals, (~~HE~~) and (A). HE with the cross through it means delete *he*. And (A) means supply *a*. The present researcher is indebted to Brown (1970, p. 44) for the reduction of Lewis's sentence to near kernel form. The signals, of course, have been supplied:

> He dipped his hands in the bichloride solution.
> He shook them. (AND ~~HE~~)

The *shake* was *quick*. (, A)
His *fingers* were *down*. (,)
His fingers were *like the fingers of a pianist above the keys*. (,)

Insertion of each sentence into the top sentence according to the present study's signal system would result in the following sentence:

> He dipped his hands in the bichloride solution and shook them, a quick shake, fingers down, like the fingers of a pianist above the keys. (Brown, 1970, p. 42)

Style has been conceived of in many ways; at times it is all-embracing, at others very narrow. Christensen (1968*a*) called it "syntax as style" (p. 572). Milic (1967) defined it as

> [the individual's] habitual and consistent selection from the expressive resources available in his language. . . . his style is the collection of his stylistic options. . . . Options or choices are not always exercised consciously. (p. 72)

McCrimmon (1968) cited the following all-encompassing definition of style given by Young and Becker (1965):

> A writer's style, we believe, is the characteristic route he takes through all the choices presented in both the writing and prewriting stages. It is the manifestation of his conception of the topic, modified by his audience, situation, and intention—what we call his "universe of discourse."

and made the following comment:

> Since all the choices cited here include those made in all three of the major classical stages, this definition subsumes invention and arrangement under style. (p. 125)

The present researcher is interested in the implications for the *teaching* of composition of considering the part of style defined above by Milic and Christensen as at least as important as invention or arrangement—style, that is, in the sense of the final syntactic choices made in the process of writing. Although lexical choices are not uncommon, the final choice made by every writer is more frequently a syntactic one. The last thing a writer usually does is to put words down on paper in a particular order. Perhaps English teachers have not sufficiently realized the desirability, indeed the necessity, of helping their students acquire the ability to put words down on paper, to manipulate syntax.

The present study's findings strongly suggest that style, rather narrowly defined as the final syntactic choices habitually made from the writer's practical repertoire of syntactic alternatives, is an important dimension of what constitutes writing ability. Whether Young and

Becker (1965) were correct in asserting that style subsumes invention and arrangement obviously cannot be answered from the findings of the present study. But it is nevertheless an interesting and an important question for rhetorical theory and practice.

Most teachers of writing either ignore or neglect the importance of syntactic manipulative ability. They certainly do not give it its proper due. And they fail to do so, perhaps, because they are concentrating on another important dimension of the writing process—that of observation and experience. Composition teachers should realize that it is not enough for a young writer to have something to say. Finally, he must be able to express it, to manipulate sentence structures in order to recapture the experience for his reader. An examination of the sentence about the small Negro girl, a sentence with all the hallmarks of a professional writer, and also of how that sentence *might* have been written should make this point clearer. Here is the "professional" example again:

> A small Negro girl develops from the sheet of glare-frosted walk, walking barefooted, her bare legs striking and recoiling from the hot cement, her feet curling in, only the outer edges touching.

After due consideration has been given to the importance of the writer's observation and experience, to the concreteness of the meticulously sequenced images, to the viewer's eye movement, all of these can be reflected in a series of sentences which are, perhaps, typical of the writing of a rather observant high school student, not that of a professional writer:

> A small Negro girl develops from the sheet of walk which is glare-frosted. She is walking barefooted. Her bare legs strike the hot cement and then they recoil from it. Her feet curl in so that only the outer edges are touching the hot cement.

In terms of observation and experience the "professional" example is no different from that of the hypothetical high school student. Both examples recreate the writer's visual experience. They carry the same experiential and observational load. The only difference between them is in the writers' handling of the syntax. It is true, of course, that the "professional" writer was able to recreate the rhythmic quality of the scene by blending syntax and image. Although there is no necessary connection between the observational experience, the ordered sequence of concrete images, and the manipulative syntactic skill of the writer, their brilliant fusion, their complementary confluence, made the difference. In the last analysis, however, the "professional" writer was able to accomplish this only through his ability to manipulate sentence struc-

ture. Teachers of writing surely ought to spend more time teaching students to be better manipulators of syntax. Intensive experience with sentence combining should help to enlarge a young writer's repertoire of syntactic alternatives and to supply him with practical options during the writing process.

The attractiveness of the sentence-combining signals in the present study lies in their simplicity, their consistency, their flexibility, and their practicality. The previous examples illustrate how simple it is both to learn to use the signals and to expand and adapt them. The elimination of the study of transformational grammar and of transformational nomenclature makes all of this possible. With the threat of grammatical failure removed, the developing writer can get on with solving sentence-structure problems and confidently face the real issue—that of blending form and idea in any given rhetorical situation.

One final comment: Although this researcher has rather strenuously urged that more attention be paid to the syntactic manipulative skill and for a more important place for "style as syntax" in the curriculum, he is merely suggesting a possible new emphasis in rhetorical instruction and is in no sense denying or even questioning the importance of the other members of the classical rhetorician's tripod, invention and arrangement. In the last analysis the question as to which of these comes first, which is more important, becomes totally irrelevant. In their essential inseparability, they are more than a tripod. Invention, arrangement and style are a trinity, one and indivisible.

Additional Suggestions for Further Research

1. It would be desirable to extend the treatment over a number of years with a population more representative of the range of ability and socioeconomic background to be found, for example, in a very large metropolitan area or in a rurally deprived area. The low ability students did very well in the present study. Would ghetto children do as well?
2. Are the growth rates attained sustainable?
3. The students seemed to enjoy the sentence-combining treatment in this study. Can interest be maintained over a number of years?
4. Can students "overlearn" sentence-combining techniques? Can they overconsolidate? Is there a ceiling on the structures they can learn? Will they be able to handle a wider variety of syntactic structures?
5. What is the connection, if any, between syntactic and cognitive

maturity? Is there a *necessary* connection between these two? *Can* and *should* syntactic maturity run ahead of cognitive maturity? Can a practical distinction be drawn between cognitive and syntactic maturity?

6. Syntactic and stylistic maturity are clearly connected. There is need for massive research into what constitutes a mature style. Hunt's (1964, 1965, 1970) and O'Donnell et al.'s (1967) normative data on syntactic maturity are based on fairly small populations. It would be useful to have, based on a larger and more representative population, normative data on the writing and speaking performance, in a variety of modes of discourse, of students from kindergarten through college.
7. Will sentence-combining practice improve reading ability?
8. Will sentence-combining practice enhance oral performance?
9. Will sentence-combining practice that involves only writing out the exercises be as successful as, or more successful than, the present study's combination of oral practice and writing practice?
10. The assessment of overall quality of the compositions by a system of forced choices of matched pairs has much to recommend it. It is economical, easy to administer, and efficient; there is no rater fatigue problem. Rating sixty compositions on a scale would have been very time consuming. This system deserves further study. No doubt it can be improved.
11. Oral and written sentence-combining practice was successful with native speakers of English. Can it be integrated into a program for teaching English as a second language?
12. One of the major problems facing teachers of a foreign language is getting students to write compositions in that language. This researcher has recently developed sentence-combining signal systems for French and Spanish and has pilot-tested them with a small number of secondary students. The results have been encouraging. This system deserves further study.

APPENDIXES

APPENDIX A

SAMPLE LESSONS AND SENTENCE-COMBINING PROBLEMS FROM THE EXPERIMENTAL GROUP'S TEXT

This appendix contains a selection of the sentence-combining problems from the first part of the experimental group's text, *Sentence-Combining*, and sample problems from eight of the more important lessons found in the second part, with instructions to the student and a sample of the cumulative problems which followed many of the lessons. Although some attempt has been made to present a representative sample of the actual progression followed in the sentence-combining lessons, there is really no substitute for a careful look at the developing, cumulative design of the student text.

In each of the examples, the A form is the sentence-combining problem confronting the student, and the B form is an acceptable student answer.

Sample Problems from the First Part of the Student Text

A. The quarterback threw the ball well yesterday. (NEG)
B. The quarterback didn't throw the ball well yesterday.

A. Lawrence Welk will turn kids on. (NEG-EVER)
B. Lawrence Welk will never turn kids on.

A. The rattler (HOW) slithered (WHERE), bit the sleeping baby (WHERE), and (HOW) disappeared.
B. The rattler quietly slithered into the tent, bit the sleeping baby on the leg, and quickly disappeared.

A. The control agents should have killed Maxwell Smart. (BY-INV)
B. Maxwell Smart should have been killed by the control agents.

A. A garbage dump is behind the restaurant. (THERE-INS)
B. There is a garbage dump behind the restaurant.

A. John was painting something on the wall. (WHAT-QUES)
B. What was John painting on the wall?

A. Someone has been copying my homework. (WHO-QUES)
B. Who has been copying my homework?

A. Some telephones are nearby. (THERE-INS + NEG + QUES)
B. Aren't there some telephones nearby?

Lesson Seven: THAT and THE FACT THAT

Julio should admit SOMETHING.
He was there.

There are three different ways to combine the two statements above into a sentence that has the same meaning: (1) Julio should admit that he was there; (2) Julio should admit he was there; and (3) Julio should admit the fact that he was there. Notice that in all three sentences the words "he was there" have been put in place of the word *SOMETHING* in the first statement. In sentence 1 the word *that* connects the given statements; in sentence 2 they are simply joined together; in sentence 3 they are joined by *the fact that.*

We are now going to practice combining statements into one sentence as was done with the three sentences above. Follow the instructions given in parentheses after the second statement. Notice that the second statement takes the place of the word *SOMETHING* in the first statement.

A. Peter noticed SOMETHING.
   There were nine golf balls in the river. (THAT)
B. Peter noticed that there were nine golf balls in the river.

A. Karen said SOMETHING.
   She wasn't going to the party. (JUST JOIN)
B. Karen said she wasn't going to the party.

A. SOMETHING should make you avoid him.
   He is an absolute nut. (THE FACT THAT)
B. The fact that he is an absolute nut should make you avoid him.

A. SOMETHING is certain.
   Human beings will survive. (THAT)
B. That human beings will survive is certain.

Lesson Eight: IT–THAT

Some of you probably wanted to write a sentence like that in the last problem in the previous section in a different way. Instead of saying, "That human beings will survive is certain," you may have preferred to say, "It is certain that human beings will survive."

In this sentence the word *It* has replaced the word *SOMETHING* and *that human beings will survive* comes after the first statement. You have simply started the sentence with *It* and then added the "that" statement later. We'll call this the (IT–THAT) instruction.

Example:

SOMETHING is true.
The world is round. (IT–THAT)
It is true that the world is round.

A. As soon as he got to the Pearly Gates, Joe told St. Peter SOMETHING had never occurred to him.
   The tires on his Jaguar might decay. (IT–THAT)
B. As soon as he got to the Pearly Gates, Joe told St. Peter it had never occurred to him that the tires on his Jaguar might decay.

A. Ever since man dragged himself out of the primeval mud and began the process of socialization, one great problem has always been SOMETHING.

He is willing to fight other men to gain power. (THAT)
B. Ever since man dragged himself out of the primeval mud and began the process of socialization, one great problem has always been that he is willing to fight other men to gain power.

A. And SOMETHING came to pass.
Cain brought an offering unto the Lord. (IT–THAT)
B. And it came to pass that Cain brought an offering unto the Lord.

A. SOMETHING made Bill believe SOMETHING.
The used car's door fell off. (THE FACT THAT)
The dealer was dishonest. (THAT)
B. The fact that the used car's door fell off made Bill believe that the dealer was dishonest.

A. SOMETHING occurred to Captain Sharp.
His men did not know SOMETHING. (IT–THAT)
They were sailing through a mined area. (THAT)
B. It occurred to Captain Sharp that his men did not know that they were sailing through a mined area.

A. SOMETHING tells the geologist SOMETHING.
The bones of fish may be found in Death Valley. (THE FACT THAT)
The region must have been under water at some time. (THAT)
B. The fact that the bones of fish may be found in Death Valley tells the geologist that the region must have been under water at some time.

Lesson Ten: WHO, WHAT, WHERE, WHEN, HOW, WHY

In previous sections you practiced combining sentences by using THAT, THE FACT THAT, and IT–THAT. In this lesson you will do something quite similar: you will combine sentences by using WHO, WHAT, WHERE, WHEN, HOW, and WHY.

Example A:

All the people wondered SOMETHING.
The music had stopped for some reason. (WHY)
All the people wondered why the music had stopped.

Notice that *for some reason* has been removed and that *why* has been inserted at the beginning of the second statement.

Example B:

SOMETHING worried the climbers.
The odd light meant something. (WHAT)
What the odd light meant worried the climbers.

Notice that *SOMETHING* has been removed and that *what* has been inserted at the beginning of the second statement.

Example C:

Most teachers have learned SOMETHING.
Students compare homework somehow. (HOW)
Most teachers have learned how students compare homework.

Notice that *somehow* has been removed and that *how* has been inserted

at the beginning of the second statement. Sentence-combining with WHO, WHERE, and WHEN would be done in similar fashion.

A. Cathy wondered SOMETHING.
The train would arrive in New York sometime. (WHEN)
B. Cathy wondered when the train would arrive in New York.

A. After the vicious murders in the downtown bank, one of the tellers gave an account of SOMETHING.
Something had happened. (WHAT)
B. After the vicious murders in the downtown bank, one of the tellers gave an account of what had happened.

A. The fish soon discovered SOMETHING.
The worm was dangling in the water for some reason. (WHY)
B. The fish soon discovered why the worm was dangling in the water.

A. Joe tried to calculate SOMETHING.
His money would buy so much food. (HOW MUCH)
B. Joe tried to calculate how much food his money would buy.

A. SOMETHING suddenly occurred to Mr. Jones.
Jim might not know SOMETHING. (IT–THAT)
Someone finds the restaurant somehow. (HOW TO)
B. It suddenly occurred to Mr. Jones that Jim might not know how to find the restaurant.

A. SOMETHING is not clear to me.
Manuel would tell you SOMETHING for some reason. (IT–WHY)
Funeral directors know SOMETHING. (THAT)
Someone solves grave problems. (HOW TO)
B. It is not clear to me why Manual would tell you that funeral directors know how to solve grave problems.

A. SOMETHING was difficult.
Jerome admitted SOMETHING. (IT–FOR–TO)
He really didn't know SOMETHING. (THAT)
The problem could be solved somehow. (HOW)
B. It was difficult for Jerome to admit that he really didn't know how the problem could be solved.

A. Because he never listens to a word the instructor is saying, SOMETHING would take hours.
Thurston learns SOMETHING. (IT–FOR–TO)
Someone puts that engine together somehow. (HOW TO)
B. Because he never listens to a word the instructor is saying, it would take hours for Thurston to learn how to put that engine together.

A. SOMETHING took real courage.
Senator Phoggbound asserted SOMETHING. (IT–FOR–TO)
He didn't care (about) SOMETHING. (THAT)
The voters thought something of him. (WHAT)
B. It took real courage for Senator Phoggbound to assert that he didn't care what the voters thought of him.

A. SOMETHING angered Miss Frump.
The girls chattered noisily. (S' + ING)
B. The girls' chattering noisily angered Miss Frump.

A. SOMETHING angered Miss Frump.
The girls chattered noisily. (S' + ~~LY~~ + ING)
B. The girls' noisy chattering angered Miss Frump.

A. SOMETHING angered Miss Frump.
The girls chattered noisily. (~~LY~~ + ING + OF)
B. The noisy chattering of the girls angered Miss Frump.

A. SOMETHING is a problem for lazy people.
Someone keeps in good health somehow. (HOW TO)
B. How to keep in good health is a problem for lazy people.

A. SOMETHING is not easy.
Mrs. Adams condoned SOMETHING. (IT–FOR–TO)
Her son was sent to Vietnam. ('S + ING)
B. It is not easy for Mrs. Adams to condone her son's being sent to Vietnam.

Lesson Seventeen: DISCOVER → DISCOVERY
ACCEPTED → ACCEPTANCE PRODUCE → PRODUCTION

We are going to practice sentence combinations which necessitate changes being made in certain word endings. For example, "Tom discovered the gold. ('S + DISCOVERY + OF)" would be written out as: "Tom's discovery of the gold. . . ." Similarly, "We failed. ('S + FAILURE)" would be written out as: "Our failure. . . ." Thus, if you were instructed to combine the following:

SOMETHING led to World War II.
The Allies punished Germany after World War I. (S' + PUNISHMENT + OF)

you would write it out like this:

The Allies' punishment of Germany after World War I led to World War II.

A. Because of numerous personality conflicts and sheer pettiness the Student Council made a mess of SOMETHING.
They formulated a set of rules for conduct. (S' + FORMULATION + OF)
B. Because of numerous personality conflicts and sheer pettiness, the Student Council made a mess of their formulation of a set of rules for conduct.

A. It would be impossible to ignore the fact that SOMETHING caused a great deal of controversy.
Simmons published the experiment. ('S + PUBLICATION + OF)
B. It would be impossible to ignore the fact that Simmons' publication of the experiment caused a great deal of controversy.

A. SOMETHING led to SOMETHING.
James Watt discovered SOMETHING. ('S + DISCOVERY)
Steam is a powerful source of energy. (THAT)
Britain established an industrial society. ('S + ING)
B. James Watt's discovery that steam is a powerful source of energy led to Britain's establishing an industrial society.

Lesson Nineteen: WHICH/THAT, WHO, and WHOM

In this lesson we'll be practicing combining sentences with WHICH/THAT, WHO, and WHOM. For example, if you were given the following:

Some of the engines were scheduled to be scrapped this year.
The saboteurs have demolished the engines. (WHICH)

you would write it like this:

Some of the engines which the saboteurs have demolished were scheduled to be scrapped this year.

Notice that *engines*, the repeated word, was replaced by *which*. You look for the repeated word when instructed to combine sentences with WHICH/THAT, WHO, and WHOM. Then you simply eliminate and substitute for one of the repeated words.

You may be given the instruction (WHICH/THAT). This simply means that you can use either *which* or *that*. You pick the one you think sounds better. Remember also that (WHICH/THAT) can mean (JUST JOIN).

A. In his letter Ralph enclosed a snapshot.
He had taken a snapshot during his visit with us. (WHICH/THAT)
B. In his letter Ralph enclosed a snapshot which he had taken during his visit with us.
or: In his letter Ralph enclosed a snapshot that he had taken during his visit with us.
or: In his letter Ralph enclosed a snapshot he had taken during his visit with us.

A. Whenever our family dines at Dino's, Grandma insists on watching the chef.
The chef tosses the pizzas high into the air. (WHO)
B. Whenever our family dines at Dino's, Grandma insists on watching the chef who tosses the pizzas high into the air.

A. Although it is usually quiet during the week, the golf course is very busy on weekends.
The golf course was completed just last year. (WHICH/THAT)
B. Although it is usually quiet during the week, the golf course that (which) was completed just last year is very busy on weekends.

A. SOMETHING is illogical.
Man believes SOMETHING. (IT–FOR–TO)
Only this tiny earth possesses the conditions. (THAT)
The conditions have made life possible. (WHICH/THAT)
B. It is illogical for man to believe that only this tiny earth possesses the conditions which have made life possible.

A. As soon as they had completed their five-mile march carrying full pack, the exhausted recruits reported to the colonel.
The colonel explained a few points. (WHO)
They had not understood a few points. (WHICH/THAT)
B. As soon as they had completed their five-mile march carrying full pack, the exhausted recruits reported to the colonel, who explained a few points (which, that) they had not understood.

Lesson Twenty: WHOSE, WHEN, WHERE, and WHY

We're now going to add instructions using WHOSE, WHEN, WHERE, and WHY. They are similar to the WHICH/THAT, WHO, and WHOM instructions, so you simply watch out for the repeated words and eliminate and substitute for them.

A. One day a girl strolled into the cafeteria.
The girl's dress looked like spun gold. (WHOSE)
B. One day a girl, whose dress looked like spun gold, strolled into the cafeteria.

A. After a wild chase through the busy downtown traffic, the young reporter was able to point out the apartment.
The gangster was hiding out in the apartment. (WHERE)
B. After a wild chase through the busy downtown traffic, the young reporter was able to point out the apartment where the gangster was hiding out.

A. The idea occurred to her at the moment.
At the moment she had all but given up hope. (WHEN)
B. The idea occurred to her at the moment when she had all but given up hope.

A. The place seemed to be enveloped in a glow.
Jill stood in the place. (WHERE)
A glow gleamed on her red hair. (WHICH/THAT)
B. The place where Jill stood seemed to be enveloped in a glow that (which) gleamed on her red hair.

A. I get nervous every time Ben goes for a swim in the ocean because he does not believe SOMETHING.
SOMETHING is possible. (THAT)
The undertow sweeps him out into deep water. (IT–FOR–TO)
B. I get nervous every time Ben goes for a swim in the ocean because he does not believe (that) it is possible for the undertow to sweep him out into deep water.

A. SOMETHING irritated Albert.
The mechanic examined the carburetor carefully. ('S + ~~LY~~ + EXAMINATION)
Albert asked SOMETHING. (WHO)
SOMETHING would take so long. (HOW LONG)
He completes SOMETHING. (IT–FOR–TO)
He inspects the whole car. ('S + INSPECTION + OF)

B. The mechanic's careful examination of the carburetor irritated Albert, who asked how long it would take for him to complete his inspection of the whole car.

Lesson Twenty-Two: Underlining as a Combining Signal

In today's lesson we are going to practice sentence-combining by eliminating repeated words and any related part of *to be* and by inserting what remains immediately after the first appearance of the repeated words. For example,

The girl suddenly began to scream in terror.
The girl was *walking through the park.*

would be written out like this:

The girl walking through the park suddenly began to scream in terror.

Notice that *the* and *girl,* the repeated words, and *was,* a form of *to be,* were eliminated. Notice also that what was left, *walking through the park,* was placed immediately after *The girl* in the first sentence.

Remember that *am, is, are, was,* and *were* are forms of *to be.*

. . . The words that have to be inserted in the sentence above will be underlined.* You simply eliminate the words that aren't inserted above. For example,

The young skater almost lost her leg in a car accident last year.
The young skater was *practicing out there on the ice.*

The young skater practicing out there on the ice almost lost her leg in a car accident last year.

You can, if it helps, put crosses through *The young skater was* in your mind's eye, but you probably won't have to. Just remember to insert the underlined words immediately after the first appearance of the repeated words.

A. Miss Jones easily smeared her attacker.
Miss Jones was *a former wrestler.*
B. Miss Jones, a former wrestler, easily smeared her attacker.

A. The governor declared in his address to the legislature that the roads will be largely paid for by taxes.
The roads are *to be built this year.*
The taxes are *on gasoline and cigarettes.*
B. The governor declared in his address to the legislature that the roads to be built this year will be largely paid for by taxes on gasoline and cigarettes.

A. Jules earned the money by pumping gas until midnight seven nights a week.

---

* The italicized words in the sentence-combining problems found in the following lessons were underlined in the experimental students' text.

Jules is *an apprentice bricklayer.*
The money was *to pay for his hiking vacation in Europe.*
B. Jules, an apprentice bricklayer, earned the money to pay for his hiking vacation in Europe by pumping gas until midnight seven nights a week.

A. The senate committee on environmental pollution did not seem to be overly impressed by the automobile industry's claim that most cars can be equipped with luxuries.
The cars are *being sold today.*
The luxuries are *to meet anyone's needs.*
B. The senate committee on environmental pollution did not seem to be overly impressed by the automobile industry's claim that most cars being sold today can be equipped with luxuries to meet anyone's needs.

A. SOMETHING is impossible.
A chef cooks meals. (IT–FOR–TO)
The chef is *working in this small kitchen.*
The meals will satisfy all customers. (WHICH/THAT)
B. It is impossible for a chef working in this small kitchen to cook meals that (which) will satisfy all customers.

A. SOMETHING angered Mr. Mulvaney.
Miss Frickert insisted SOMETHING. ('S + ING)
There were spooks in the house. (THAT)
She had just rented the house. (WHICH/THAT)
Mr. Mulvaney is *the policeman on our block.*
B. Miss Frickert's insisting that there were spooks in the house (which, that) she had just rented angered Mr. Mulvaney, the policeman on our block.

A. SOMETHING irritated the men.
Connie constantly chattered. ('S + ~~LY~~ + ING)
The chattering kept the hunters from hearing something. (WHICH/THAT)
The dogs were running someplace. (WHERE)
The men swore SOMETHING. (WHO)
They would never take her hunting again. (THAT)
B. Connie's constant chattering, which kept the hunters from hearing where the dogs were running, irritated the men, who swore (that) they would never take her hunting again.

Lesson Twenty-Four: The Underlining Signal Continued

In the previous section you practiced inserting the underlined words immediately after the first appearance of the repeated words. In this lesson there will be single words underlined. Most of the time, you will insert these single words in front of the first appearance of the repeated words. For example,

Most Latin Americans prefer soccer to bullfighting.
The Latin Americans are *youthful.*

would be written out as:

Most youthful Latin Americans prefer soccer to bullfighting.

Notice that *the* and *Latin Americans,* the repeated words, and *was,* a form of *to be,* were eliminated. Notice also that *youthful,* which was underlined, was placed in front of *Latin Americans.* Let's look at another example.

We saw a fourteen-year-old girl selling heroin to four twelve-year-olds at school.
The girl was *cruelly undernourished.*

would be written out as,

We saw a cruelly undernourished fourteen-year-old girl selling heroin to four twelve-year-olds at school.

Put the underlined phrases where you think they fit best, before or after.

A. All the English students loved their teacher.
The students were *cool.*
The teacher was *charming.*
B. All the cool English students loved their charming teacher.

A. The cabinet official, who obviously knew nothing about economics, declared that a budget is of overwhelming importance.
The budget has been *balanced.*
B. The cabinet official, who obviously knew nothing about economics, declared that a balanced budget is of overwhelming importance.

A. The alleys were littered with bottles and garbage.
The alleys were *between the apartment buildings.*
The apartment buildings were *dismal.*
The bottles were *broken.*
The garbage was *rotting.*
B. The alleys between the dismal apartment buildings were littered with broken bottles and rotting garbage.

A. The explorers saw formations.
The formations were *glistening.*
The formations were *black.*
The formations were *rock.*
The formations were *rising hundreds of feet into the air.*
The formations were *one of Asia's greatest wonders.*
B. The explorers saw glistening black rock formations rising hundreds of feet into the air, one of Asia's greatest wonders.

A. Some teachers often hesitate to give students answers to those questions.
The teachers are *rather timid.*
The answers are *frank.*
The answers are *personal.*
The questions are *basic.*
The questions disturb us all as we try to understand our lives. (WHICH/THAT)
B. Some rather timid teachers often hesitate to give students frank, per-

sonal answers to those basic questions which (that) disturb us all as we try to understand our lives.

A. The office building towered above the apartment houses.
The building was *gleaming*.
The building was *new*.
The building was *rising high into the sky*.
The houses were *decrepit*.
The houses were *brick*.
The houses were *in the slums*.
The slums surrounded this symbol of prosperity. (WHICH/THAT)
The prosperity was *universal*.

B. The gleaming new office building, rising high into the sky, towered above the decrepit, brick apartment houses in the slums which (that) surround this symbol of universal prosperity.

A. A girl tightly held the hand of her mother.
The girl was *pale*.
The girl was *nervous*.
The girl was *about six years old*.
The girl was apparently going to school for the first time. (WHO)
Her mother was *smiling*.
Her mother calmly encouraged her. (WHO)

B. A pale, nervous girl about six years old, who was apparently going to school for the first time, tightly held the hand of her smiling mother, who calmly encouraged her.

A. The heron tensed its wings for the plunge.
The heron was *princely*.
The heron was *perched high on a ledge*.
The ledge was *rocky*.
The ledge's height enabled the bird to survey the waters. (WHOSE)
The waters were *swirling*.
The waters were *blue-white*.
The waters were below on three sides. (WHICH/THAT)
The plunge would be *spectacular*.
The plunge was *soon to be triggered by a school of fish*.
The school of fish were *fast approaching*.

B. The princely heron, perched high on a rocky ledge whose height enabled the bird to survey the swirling blue-white waters that were below on three sides, tensed its wings for the spectacular plunge soon to be triggered by a fast approaching school of fish.

A. The soldiers realized SOMETHING.
The soldiers were *weary*.
The soldiers were *battle-scarred*.
The soldiers were *in the thick of that struggle*.
The struggle was *bloody*.
The struggle was *desperate*.
The struggle was *hand-to-hand*.
The struggle took the lives of so many men. (WHICH/THAT)
The men were *fine*.

The men were *young.*
SOMETHING would be difficult. (THAT)
They would survive the mistakes of generals. (IT–FOR–TO)
The mistakes were *costly.*
The generals were *inexperienced.*
The generals relied on tactics. (WHO)
The tactics were *outmoded.*
The tactics were *military.*
The tactics simply did not fit the realities of warfare. (WHICH/THAT)
The warfare was *modern.*

B. The weary battle-scarred soldiers in the thick of that bloody, desperate, hand-to-hand struggle, which took the lives of so many fine young men, realized that it would be difficult for them to survive the costly mistakes of inexperienced generals who relied on outmoded military tactics that simply did not fit the realities of modern warfare.

A. Mr. Lippman has suggested SOMETHING.
Mr. Lippman is *a noted columnist.*
The President has committed us to a war. (THAT)
The war is *in Asia.*
The war is *for an objective.*
The objective is *unattainable.*
The objective is SOMETHING.
Someone creates a government. (, THE + CREATION + OF)
The government is *secure.*
The government is *free.*
The government is *pro-American.*
The government is *accepted and supported by the people.*

B. Mr. Lippman, a noted columnist, has suggested that the President has committeed us to a war in Asia for an unattainable objective, the creation of a secure, free, pro-American government accepted and supported by the people.

A. The seventh graders could not understand SOMETHING.
The seventh graders had worked hard on their assignments. (WHO)
The assignments were *English.*
They had worked *all year.*
Their teacher had assigned two reports for some reason. (WHY)
The reports were *written.*
The reports were *per week.*
The reports were *on some novels.*
The novels were *boring.*
The novels would make SOMETHING impossible. (WHICH/THAT)
They would fully enjoy their summer vacations. (IT–FOR–TO)

B. The seventh graders who had worked hard on their English assignments all year could not understand why their teacher had assigned two written reports per week on some boring novels that would make it impossible for them to fully enjoy their summer vacations.

APPENDIX B

SAMPLE LESSONS AND PROBLEMS FROM AN EXPANDED VERSION OF THE SENTENCE-COMBINING SYSTEM USED IN THIS STUDY

In this study it was suggested that the sentence-combining system could be expanded to include a larger number of the syntactic structures of English. This appendix contains a number of lessons introducing sentence-combining signals which have been added since the study was completed. The lessons were pilot-tested with several hundred high school students and their teachers. The actual lessons and sentence problems were taken from a sentence-combining text which this researcher has completed for Ginn and Company.

Please note that these lessons and sentence-combining signals were *not* used in this study.

The addition of these signals makes it possible to construct sentence-building problems encompassing most of the syntactic structures in the language. These additional signals have been included in the hope that English teachers will use this expanded sentence-combining system in their classes. An hour or two of practice is all that is needed to become fairly adept at reducing selected sentences to near-kernel form and adding the appropriate signals.

Section One: X-ing Out

Although we'll be concentrating on having you practice longer sentences than you're used to writing, we are *not* saying that all your own sentences should be long. You're going to learn not only to add but to delete, to get rid of the superfluous and, hopefully, to achieve efficiency and clarity. You'll learn to juxtapose long and short sentences to make them both more effective. A composition consisting solely of long sentences is likely to be as dull and immature as one made up exclusively of short sentences. In this book, as in life, variety is the spice.

Your first sentence-combining practice involves three simple operations. Given a series of statements, you get rid of certain words, add the appropriate commas, and add only one word: *and.* Observe how this works. Given the following three base sentences,

> The blunt nose of the Hindenburg bobbed up.
> The blunt nose hung a moment in the air.
> Then it crumpled toward the field. (AND)

you could write,

> The blunt nose of the Hindenburg bobbed up, hung a moment in the air, and then crumpled toward the field.

Three things happened:

1. "The blunt nose" and "it" were deleted.
2. Commas were added.
3. An *and* was introduced just before the last base sentence.

In the following example you'll be given signals to help you decide which words to delete and where to put commas.

> ~~She~~ means delete "She."
> The comma signal (,) means put a comma at the beginning of this base sentence.
> (,AND) means put a comma followed by an *and* at the beginning of this base sentence.

Given the following three base sentences,

> Helen raised her pistol.
> ~~She~~ took careful aim. (,)
> ~~She~~ squeezed off five rapid shots to the center of the target. (, AND)

you would write them out like this:

> Helen raised her pistol, took careful aim, and squeezed off five rapid shots to the center of the target.

Each of the following sets of base sentences is to be written as just one sentence. . . . In the first two problems you have been supplied with AND, X-ing out, and comma signals. In the remaining problems decide for yourself which words are to be X-ed out, where to put commas, and whether an *and* is needed.

A. The pitcher looked up intently.
~~The pitcher~~ glanced at first base. (,)
Then ~~he~~ threw a hanging curve which the batter knocked out of the stadium. (, AND)
B. The pitcher looked up intently, glanced at first base, and then threw a hanging curve which the batter knocked out of the stadium.

A. Carlos smoked their cigarettes.
~~He~~ lounged with his feet on their couch. (,)
~~He~~ occasionally took Juanita places in their car. (, AND)
B. Carlos smoked their cigarettes, lounged with his feet on their couch, and occasionally took Juanita places in their car.

A. The Hindenburg burst like a bomb.
It crashed in flames with ninety-seven persons on board.
B. The Hindenburg burst like a bomb and crashed in flames with ninety-seven persons on board.

A. When John didn't turn up for their date, Sally walked down to the bridge.
Sally climbed up on the rails.
She did a neat dive into the river.

B. When John didn't turn up for their date, Sally walked down to the bridge, climbed up on the rails, and did a neat dive into the river.

A. The Hindenburg rode out a severe storm along the Atlantic coast.
The Hindenburg glided safely through and around angry forks of lightning.
Then the Hindenburg, with a safe landing in her grasp, plummeted to the ground.
B. The Hindenburg rode out a severe storm along the Atlantic coast, glided safely through and around angry forks of lightning, and then, with a safe landing in her grasp, plummeted to the ground.

A. Fearless Fred dashed into the room.
He dived at the dastardly robber.
He missed.
He went sailing out of the five-story window.
B. Fearless Fred dashed into the room, dived at the dastardly robber, missed, and went sailing out of the five-story window.

A. They walked on.
They were looking at the stars.
They were talking about them.
They were ignoring the deserted look the cottages wore.
They were pretending not to see the cars that passed them.
B. They walked on, looking at the stars, talking about them, ignoring the deserted look the cottages wore, pretending not to see the cars that passed them.

Section Two: Making the Connection

One of the simplest ways of combining two sentences is to put them back-to-back with a connecting word between. The connecting word establishes a relationship between the two base sentences, a relationship that might be hard to establish by any other means. The relationships are mainly of (1) cause-effect, (2) time, and (3) similarity or difference. Here is an example of each kind of relationship:

1. They were happy because their team won.
2. They were happy when their team won.
3. They were happy, but we were miserable.

Sometimes you can establish a relationship without specifying which kind. The semicolon (;) is the "connection":

4. They were happy; their team won.

Some connecting words are *as soon as, just when, after, before, although, if, since*.

A brief look at the combining mechanics is all you need before plunging into the practice. When one of the connecting words appears as an instruction, attach it to the beginning of its base sentence. Then attach the result to the beginning of the following sentence or the end of the previous sentence.

A. The soldiers came home.
The war ended. (WHEN)
B. The soldiers came home when the war ended.

A. The war ended. (WHEN)
The soldiers came home.
B. When the war ended, the soldiers came home.

A. I don't get there by midnight. (IF)
Come looking for me.
I'll be in trouble. (;)
B. If I don't get there by midnight, come looking for me; I'll be in trouble.

A. He always quits.
You need him. (JUST WHEN)
B. He always quits just when you need him.

A. You overcome your fear of the water. (ONCE)
Learning to swim becomes a matter of patience and practice.
B. Once you overcome your fear of the water, learning to swim becomes a matter of patience and practice.

A. Night came. (WHEN)
We sat huddled in blankets.
The blankets were *thick* and *woolly*.
It was time to turn in for the night. (LONG BEFORE)
B. When night came, we sat huddled in thick, woolly blankets long before it was time to turn in for the night.

Section Three: (ING) and (WITH)

Some really effective sentences can be constructed by changing a word to its *-ing* form or by using *with* as a "connector." Notice how the (ING) instruction works:

A. Joe *burst through the line.* (ING)
Joe forced the quarterback to eat the ball on fourth down.
B. Bursting through the line, Joe forced the quarterback to eat the ball on fourth down.

The (ING) instruction causes *burst* to become *bursting*, and the italics are a reminder to get rid of the *Joe* in that sentence. Now try one of your own:

A. The angry crowd *fell on the assassin.* (ING)
The angry crowd tore him limb from limb.

The (WITH) instruction does one of two things, depending on the kind of sentence it follows. Look at these examples:

A. She was a sensuous looking beauty.
She had long auburn hair. (WITH)
B. She was a sensuous looking beauty with long auburn hair.

A. Her car was in a four-wheel drift. (WITH)
She counter-steered and went on to take the lead.

B. With her car in a four-wheel drift, she counter-steered and went on to take the lead.

In the first example, (WITH) "kicks out" the *had* and the *she* and settles at the beginning of the sentence. In the second example, (WITH) simply eliminates the *was* and settles at the beginning of the sentence.

A. The slave *cried out for mercy.* (ING)
The slave threw himself at the sultan's feet.
The slave had been caught in the harem. (WHO)
B. Crying out for mercy, the slave, who had been caught in the harem, threw himself at the sultan's feet.

A. It was a wild wet day.
The wind was slapping at your face. (WITH)
The wind *chilled you through and through.* (ING)
B. It was a wild wet day with the wind slapping at your face, chilling you through and through.

A. Alex was *lonely.*
Alex was *disillusioned.*
Alex was *bitter.*
Alex shuffled into the bus station.
His shoulders were *bowed.*
His suitcase was *heavy in his hand.*
B. Lonely, disillusioned, bitter, Alex shuffled into the bus station, his shoulders bowed, his suitcase heavy in his hand.

A. Robert was *dedicated.*
Robert was *honest.* (AND)
Robert was doomed to failure in a society.
The society sneered at dedication. (THAT)
The society *refused to acknowledge selfless commitment.* (ING)
B. Dedicated and honest, Robert was doomed to failure in a society that sneered at dedication, refusing to acknowledge selfless commitment.

A. The deer *sensed danger.* (ING)
The deer lifted its head.
Its *ears* were *stiff and straight.*
Its *body* was *tense.*
It was *ready to explode into motion at the slightest sound.*
B. Sensing danger, the deer lifted its head, ears stiff and straight, body tense, ready to explode into motion at the slightest sound.

A. You got beyond those pious utterances about his concern for the weak and oppressed. (WHEN)
You realized SOMETHING.
He was quite simply an egomaniac. (THAT)
He had no other concern but his own selfish ambition. (WITH)
B. When you got beyond those pious utterances about his concern for the weak and oppressed, you realized that he was quite simply an egomaniac with no other concern but his own selfish ambition.

A. They were *hand in hand.*
They walked on in silence.
The wind stirred the moist, warm air. (ING)
The tide swept rhythmically over their bare feet. (ING)
The sand was *cool and liquid on their toes.*

B. Hand in hand, they walked on in silence, the wind stirring the moist, warm air, the tide sweeping rhythmically over their bare feet, the sand cool and liquid on their toes.

A. Julia stood at the edge of the cliff.
She *looked down on their upturned, nickel-sized faces by the side of the tidal pool.* (ING)
She *wished she had ignored the dare.* (ING)
She *felt trapped.* (ING)
Yet she *knew SOMETHING.* (ING)
She couldn't back down. (THAT)

B. Julia stood at the edge of the cliff, looking down on their upturned, nickel-sized faces by the side of the tidal pool, wishing she had ignored the dare, feeling trapped, yet knowing that she couldn't back down.

A. The gas station attendant stumbled out of his shack.
He was *an emaciated looking fellow.*
He had white hair and skin the color of an old saddle. (WITH)
He *stood scowling at us.* (AND)
His chin was thrust forward. (WITH)
His eyes were *blazing.*

B. The gas station attendant, an emaciated looking fellow with white hair and skin the color of an old saddle, stumbled out of the shack and stood scowling at us, with his chin thrust forward, his eyes blazing.

Section Four: Colon and Dash

The colon (:) is favored by many professional writers and is a very useful writing device. For example, in *Andersonville* MacKinlay Kantor has a sentence whose bases look like this:

A. He had managed to buy also a coral necklace for his small daughter.
It was *a coral necklace naturally.* (:)
Her name was Coralie. (SINCE)

The original sentence is:

B. He had managed to buy also a coral necklace for his small daughter: a coral necklace naturally, since her name was Coralie.

Notice that the colon (:) went to the front of its base sentence and that the italics shows which part of the sentence is to be retained. Here is a sentence, broken down into bases, from John Updike's story, "Who Made Yellow Roses Yellow?":

A. He pushed back the chair a few feet.
A full view of himself was available in the tilted mirror. (SO)

He was *a tall, narrow-skulled, smooth-cheeked youth.* (:)
The youth was *tightly dressed in darkest gray.*

And here is Updike's original sentence, "reconstituted" by following the various instructions:

B. He pushed back the chair a few feet, so a full view of himself was available in the tilted mirror: a tall, narrow-skulled, smooth-cheeked youth, tightly dressed in darkest gray.

Notice, again, that the colon (:) went to the front of its base sentence.

Updike also makes effective use of the dash (—). In the same story he wrote a sentence that looks like this when broken down into base sentences:

A. It's a terrific image.
The image is of *this perceptive man.* (—)
The man is *caged in his own weak character.*

"Reconstituted," the sentence looks like this:

B. It's a terrific image—this perceptive man caged in his own weak character.

Notice that the dash (—) went to the front of its base sentence.

Writers often use two dashes to separate what they insert into a sentence from the rest of that sentence. The first dash will go at the front of its base sentence. The second dash is often written immediately after its base sentence; the instruction for this will be three dots and a dash (. . . —). The three dots are telling you first to write in the base sentence (changing it according to any other instructions) and then to put in the dash. Here's an example from Updike:

A. George read into each irregular incident possible financial loss.
The incident could be *a greeting on the subway.* (—)
The incident could be *an unscheduled knock on the door.* (. . . —)

Updike's original looks like this:

B. George read into each irregular incident—a greeting on the subway, an unscheduled knock on the door—possible financial loss.

Now try combining sentences using the colon and dash instructions. All the sentences in this practice have been taken from the works of modern writers. Remember that (WHICH) and (THAT) can and often do mean (JOIN).

A. And we have SOMETHING.
They sorely need something. (WHAT)
They need *a new sense of life's possibilities.* (:)
(James Baldwin)
B. And we have what they sorely need: a new sense of life's possibilities.

A. Different as they were.
They were different *in background.* (—)
They were different *in personality.*
They were different *in underlying aspiration.* (. . . —)

These two great soldiers had much in common.
(Bruce Catton)

B. Different as they were—in background, in personality, in underlying aspiration—these two great soldiers had much in common.

A. I had never seen a man beaten.
He had been beaten. (AS)
He was *this mountain of a man.* (—)
He had died in the battle. (WHO)
He had been fighting the battle for forty-six years. (WHICH)
(Jesse Stuart)

B. I had never seen a man beaten as he had been beaten—this mountain of a man, who had died in the battle which he had been fighting for forty-six years.

A. The crimes have changed in rapid succession.
The Jews have been charged with the crimes in the course of history. (WHICH)
They were *crimes.* (—)
The crimes were to justify the atrocities. (WHICH)
The atrocities were *perpetrated against them.* (. . . —)
(Albert Einstein)

B. The crimes which the Jews have been charged with in the course of history—crimes which were to justify the atrocities perpetrated against them—have changed in rapid succession.

A. She studied him.
She answered. (BEFORE)
He was *tall.* (:)
He was *not too big or heavy.*
He was *black.* (AND)
(Shirley Ann Grau)

B. She studied him before she answered: tall, not too big or heavy, and black.

A. Open and peaceful competition is something else again.
The competition is *for prestige.* (—)
The competition is *for markets.*
The competition is *for scientific achievement.*
The competition is *even for men's minds.* (. . . —)
(John F. Kennedy)

B. Open and peaceful competition—for prestige, for markets, for scientific achievements, even for men's minds—is something else again.

A. Never shall I forget the deep singing of the men at the drum.
The singing of the men at the drum was *swelling and sinking.*
It was *the deepest sound I have ever heard in all my life.*
It was *deeper than thunder.*
It was *deeper than the sound of the Pacific Ocean.*
It was *deeper than the roar of a deep waterfall.*
It was *the wonderful deep sound of man.* (:)
Man was *calling to the unspeakable depths.*
(D. H. Lawrence)

B. Never shall I forget the deep singing of the men at the drum, swelling and sinking, the deepest sound I have heard in all my life, deeper than thunder, deeper than the sound of the Pacific Ocean, deeper than the roar of a deep waterfall: the wonderful deep sound of man calling to the unspeakable depths.

APPENDIX C

COMPOSITION EVALUATION ASSIGNMENTS

Assignment A-1

We all enjoy an unusual story, especially the kind which holds our interest and makes us wonder what will happen next. Below are listed four titles. Choose the one which seems most interesting to you, and write a story that fits the title and is mysterious or strange. Use your imagination to fill in the details, and make sure you tell the complete story, from beginning to end. Try to make it sound as if it really happened.

Stranded in a Ghost Town
The Thing that Wouldn't Die
No Ordinary Forest!
The Strangest Day Ever

Instructions. Plan your story so that it is as clear as possible. Use the back of this paper to jot down and organize your ideas. Then write your story on the *lined paper*. You will probably have written from seven to fifteen sentences by the time you finish. You have until the end of the period to complete the story.

Assignment B-1

Unusual stories are enjoyable. We all like stories which hold our attention and make us wonder what is coming next. Choose *one* title from the four listed below, the one which is most interesting to you. Now write a mysterious or strange story which fits that title. Fill in the details from your own imagination, and be sure to tell the whole story, from start to finish. Try to make it sound as if it really happened.

Creature From the Lake
The Old Woman in the Fog
What an Unusual Day!
Lost on Evil Island

Instructions: Plan your story so that it is as clear as possible. Use the back of this paper to jot down and organize your ideas. Then write your story on the *lined paper*. You will probably have written from seven to fifteen sentences by the time you finish. You have until the end of the period to complete the story.

Assignment A-2

All of us like to remember the special places we've visited and the exciting

things we've done. In order to help our friends share these experiences, we should try to tell them things that will help them make a mental picture of the scene: the colors, the motions, the sounds, the smells, and the feelings which made the experience mean so much to us. Choose *one* of the four topics listed below. Use your imagination and your writing skill to create a lifelike word picture of such a scene, and the impressions it made on you. Concentrate on describing the sights, sounds, and smells, rather than merely telling what happened.

The Dinner Table at Thanksgiving
A Rainy Night at the Football Game
Watching a Building Burn Down
The Amusement Park at Night

Instructions. Plan your story so that it is as clear as possible. Use the back of this paper to jot down and organize your ideas. Then write your story on the *lined paper*. You will probably have written from seven to fifteen sentences by the time you finish. You have until the end of the period to complete the story.

Assignment B-2

All of us like to remember the special places we've visited and the exciting things we've done. In order to help our friends share these experiences, we should try to tell them things that will help them make a mental picture of the scene: the colors, the motions, the sounds, the smells, and the feelings which made the experience mean so much to us. Choose *one* of the four topics listed below. Use your imagination and your writing skill to create a lifelike word picture of such a scene, and the impressions it made on you. Concentrate on describing the sights, sounds, and smells, rather than merely telling what happened.

An Afternoon at a Fair in Autumn
A Cook-Out at the Shore
After a Bad Storm
Halloween Night

Instructions. Plan your story so that it is as clear as possible. Use the back of this paper to jot down and organize your ideas. Then write your story on the *lined paper*. You will probably have written from seven to fifteen sentences by the time you finish. You have until the end of the period to complete the story.

Assignment A-3

Whenever we feel strongly about something, we often try to persuade others to think as we do or to do what we want them to do. We usually try to think of as many good reasons as possible to persuade them to do it. It's also a good idea for us to show that the reasons against doing it are not as convincing. Now choose one of the situations listed below, and write a composition in which you try to persuade the person named to do what you want him to do.

A lady in your neighborhood is not sure if she should have you babysit. Convince her that you can do the job.

Persuade your parents to raise your allowance by a certain amount.

Convince your parents that you should be allowed to go to the high school dance.

Assignment B-3

Whenever we feel strongly about something, we often try to persuade others to think as we do or to do what we want them to do. We usually try to think of as many good reasons as possible to persuade them to do it. It's also a good idea for us to show that the reasons against doing it are not as convincing. Now choose one of the situations listed below, and write a composition in which you try to persuade the person named to do what you want him to do.

Imagine your parents have won a competition, and the prize is a two week vacation, all expenses paid. Persuade them to go to the place of your choice.

Persuade a teacher to limit homework to one night a week.

Convince your parents that you should be allowed to decide how you dress.

Assignment A-4

In Davy Crockett's time transportation was a very important part of people's lives. People in those days walked a great deal, but they also made use of horses, mules, coaches, canal boats and ships to carry men and goods from place to place. Imagine that a boy (or girl) your age living in those days accidentally entered the fourth dimension and landed in Tallahassee this week. Naturally he would be fascinated by the changes in the means of transportation that have taken place since the frontier days. *Write a report,* telling him of the many new means of transportation that have been invented since his day. Tell him how they work, what they can do, and so on. Try to answer any questions you think he might ask.

Assignment B-4

In Jim Bowie's time most frontier people's homes were log cabins, lit by oil lamps or candles. They had no running water and the few kitchen utensils they had were usually crude and simple. Imagine that a boy (or girl) your age living in those days accidentally entered the fourth dimension and landed in Tallahassee this week. Naturally he would be fascinated by the many changes that have taken place in homes since the frontier days. *Write a report,* telling him of the developments in the home since his day. Mention some of the many home appliances and gadgets that have been invented since Jim Bowie's time. Tell him how they work, what they can do, and so on. Try to answer any questions you think he might ask.

Assignment A-5

All of us have often been to particular places that made us feel good to be there. Sometimes, if we go back again to the same place, we have the same good feeling all over again, because of how the place looked or felt, or because of what we did or what happened to us the last time.

Choose *one* of the places listed below which makes you feel *good* when you go there. Try to describe the place (what you saw, how it made you feel, the people who were there) so that a teenage friend will understand why you feel good about the place.

Saturday afternoon at the movies
An ice cream parlor
Sunday morning at Church
A favorite quiet place
A backyard swimming pool

Assignment B-5

All of us have been to particular places that made us feel *bad* to be there. Sometimes, if we go back again to the same place, we have the same bad feeling all over again, because of what we did or what happened to us the last time.

Choose *one* of the places listed below which makes you feel *bad* when you go there. Try to describe the place (what you saw, how it made you feel, the people who were there) so that a teenage friend will understand why you feel bad about the place.

A doctor's or dentist's office
The school principal's office
A new school on the first day
A traffic jam on a hot afternoon
Your room on a cold winter morning,
when the heating system has broken down.

REFERENCES

Anderson, C. C. 1960. The new STEP essay test as a measure of composition ability. *Educational and Psychological Measurement* 20:95–102.

Anderson, J. E. 1937. An evaluation of various indices of linguistic development. *Child Development* 8:62–68.

Bateman, D. R., and Zidonis, F. J. 1964. *The effect of a knowledge of generative grammar upon the growth of language complexity*. Columbus: The Ohio State University. U.S. Office of Education Cooperative Research Project 1746.

———. 1966. *The effect of a study of transformational grammar on the writing of ninth and tenth graders*. Research Report no. 6. Urbana, Ill.: National Council of Teachers of English.

Braddock, R., Lloyd-Jones, R., and Schoer, L. 1963. *Research in written composition*. Urbana, Ill.: National Council of Teachers of English.

Brown, J. A. 1970. Concepts, kernels, and composition. *South Atlantic Bulletin* 35(2): 41–45.

Buxton, E. W. 1958. An experiment to test the effects of writing frequency and guided practice upon students' skill in written expression. Unpublished doctoral dissertation, Stanford University.

Campbell, D. T., and Stanley, J. C. 1968. *Experimental and quasi-experimental designs for research*. Chicago: Rand-McNally.

Carroll, J. B. 1960. Language development. In *Encyclopedia of educational research*, ed. C. W. Harris. New York: The Macmillan Company.

Chomsky, N. 1957. *Syntactic structures*. The Hague: Mouton Publishers.

———. 1965. *Aspects of the theory of syntax*. Cambridge: The M.I.T. Press.

Chotlos, J. W. 1944. Studies in language behavior IV: A statistical and comparative analysis of individual written language samples. *Psychological Monographs* 56:77–111.

Christensen, F. 1967. *Notes toward a new rhetoric*. New York: Harper & Row.

———. 1968*b*. *The Christensen rhetoric program*. New York: Harper & Row.

———. 1968*a*. The problem of defining a mature style. *English Journal* 57:572–579.

Cochran, W. G. 1950. The comparison of percentages in matched samples. *Biometrika* 37:256–266.

Diedrich, P. E., French, S. W., and Carlton, S. T. 1961. *Factors in judgments of writing ability*. Research Bulletin RB–61–15. Princeton: Educational Testing Service.

Dixon, W. J., and Massey, F. K. 1969. *Introduction to statistical analysis*. New York: McGraw-Hill.

Erwin, S., and Miller, W. R. 1963. Language development. In *Child Psy-*

chology, ed. H. W. Stevenson. Chicago: National Society for the Study of Education, pp. 108–143.

Frogner, E. 1933. Problems of sentence structure in pupils' themes. *English Journal* 22:742–749.

Gleason, H. A. 1962. What is English? *College Composition and Communication* 13(3):1–10.

Griffin, W. J. 1967. Developing syntactic control in seventh grade writing through audio-lingual drill on transformations. Paper read at American Educational Research Association Convention, New York.

Harrell, L. E., Jr. 1957. A comparison of the development of oral and written language in school-age children. *Monographs of the Society for Research in Child Development*, vol. 22, no. 3.

Heider, F., and Heider, G. M. 1940. A comparison of sentence structure of deaf and hearing children. *Psychological Monographs* 52(1):42–103.

Hunt, K. W. 1964. Differences in grammatical structures written at three grade levels, the structures to be analyzed by transformational methods. Tallahassee: Florida State University. U.S. Office of Education Cooperative Research Project 1998.

———. 1965. *Grammatical structures written at three grade levels.* Research Report no. 3. Urbana, Ill.: National Council of Teachers of English.

———. 1970. Syntactic maturity in schoolchildren and adults. *Monographs of the Society for Research in Child Development*, serial no. 134, vol. 35, no. 3.

Hunt, K. W., and O'Donnell, R. 1970. An elementary school curriculum to develop better writing skills. Tallahassee: Florida State University. U.S. Office of Education Grant No. 4–9–08–903–0042–010.

Johnson, S. T. 1969. Some tentative strictures on generative rhetoric. *College English* 31:155–165.

Kincaid, G. L. 1953. Some factors affecting variations in the quality of students' writing. Unpublished doctoral dissertation, Michigan State University.

Lees, R. B. 1960. The grammar of English nominalizations. *International Journal of American Linguistics*, vol. 26, no. 3, part 2.

———. 1961. Grammatical analysis of the English comparative construction. *Word* 17(2):171–185.

Loban, W. D. 1961. Language ability in the middle grades of elementary school. U.S. Office of Education Cooperative Research Project SAE 7287.

———. 1963. *The language of elementary school children.* Research Report no. 1. Urbana, Ill.: National Council of Teachers of English.

McCarthy, D. 1954. Language development in children. In *Manual of child psychology*, ed. L. Carmichael. New York: John Wiley & Sons.

McCrimmon, J. M. 1969. Will the new rhetorics product new emphases in the composition class? *College Composition and Communication* 20:124–130.

Meckel, H. C. 1963. Research on teaching composition and literature. In *Handbook of research on teaching*, ed. N. L. Gage. Chicago: Rand-McNally.

Mellon, J. C. 1965*a*. Methodological problems in relating grammar instruction to writing ability: a critical review of selected research. Unpublished qualifying paper, Graduate School of Education, Harvard University.

———. 1965*b*. *Our sentences and their grammar: a modern grammar of English, grade seven.* Graduate School of Education, Harvard University.

———. 1967. Transformational sentence-combining: a method for enhancing the development of syntactic fluency in English composition. Cambridge: Harvard University. U.S. Office of Education Cooperative Research Project 5–8418.

———. 1969. *Transformational sentence-combining: a method for enhancing the development of syntactic fluency in English composition.* Research Report no. 10. Urbana, Ill.: National Council of Teachers of English.

Milic, L. T. 1967. Against the typology of styles. In *Rhetoric: theories for application,* ed. R. M. Gorrell. Urbana, Ill.: National Council of Teachers of English.

Miller, G. A. 1956. The magic number seven, plus or minus two. *Psychological Review* 63:81–97.

Miller, B. D., and Ney, J. W. 1968. The effect of systematic oral exercises on the writing of fourth-grade students. *Research in English* 2:44–61.

Moffett, J. W. 1968*a*. *A student-centered language arts curriculum, grades K–13.* Boston: Houghton-Mifflin.

———. 1968*b*. *Teaching the universe of discourse.* Boston: Houghton-Mifflin.

Ney, J. W. 1966. Applied linguistics in the seventh grade. *English Journal* 55:895–897.

———. 1968. Conditioning syntactic performance of children at varying grade levels by audio-lingual drills on transformation. Paper read at American Educational Research Association Convention, Chicago.

Noyes, E. S. 1963. Essay and objective tests in English. *College Board Review,* no. 49.

O'Donnell, R. C., Griffin, W. J., and Norris, R. C. 1967. *Syntax of kindergarten and elementary school children: a transformational analysis.* Research Report no. 8. Urbana, Ill.: National Council of Teacher of English.

Postman, N. 1967. Linguistics and the pursuit of relevance. *English Journal* 56:1160–1165.

Raub, D. K. 1966. The audio-lingual drill technique: an approach to teaching composition. Unpublished M.A. thesis, George Peabody College for Teachers.

Seegers, J. C. 1933. Form of discourse and sentence structure. *Elementary English Review* 10:51–54.

Stalnaker, J. M. 1934. The construction and results of a twelve-hour test in English composition. *School and Society* 39:218–224.

Starring, R. W. 1952. A study of ratings of comprehensive examination themes when certain elements are weakened. Unpublished doctoral dissertation, Michigan State College.

Young, R. E., and Becker, A. L. 1965. Toward a modern theory of rhetoric: a tagmemic contribution. *Harvard Educational Review* 35:64–66.